Scanning the Skies

Scanning the Skies

A History of Tornado Forecasting

MARLENE BRADFORD

University of Oklahoma Press : Norman

This book is published with the generous assistance of The McCasland Foundation, Duncan, Oklahoma.

Library of Congress Cataloging-in-Publication Data

Bradford, Marlene.
Scanning the skies : a history of tornado forecasting / Marlene Bradford.
p. cm.
Includes bibliographical references and index.
ISBN 0-8061-3302-3 (alk. paper)
1. Tornadoes—United States. 2. Weather forecasting—United States.
3. Tornado warning systems—United States. I. Title.

QC955 .B72 2001
551.64′53′0973—dc21 00-059979

The paper in this book meets the guidelines for permanence and durability of the Committee on Production Guidelines for Book Longevity of the Council on Library Resources, Inc. ∞

1 2 3 4 5 6 7 8 9 10

To my mother,
who instilled in me a passion for weather

Contents

ILLUSTRATIONS

Tables

Preface

Tornadoes fascinate me. I have always lived in tornado country, and although I have experienced only one small tornado, I have great respect for their power. The destruction these monsters can unleash should motivate those in a tornado's path to take whatever actions are necessary to save their lives. One essential component of saving lives from tornadoes is a tornado forecast, or, in today's terminology, a tornado watch.

My first memory of tornado watches was during my school days in Fort Worth, Texas, when frequently in the spring the principal would dismiss classes early if a tornado watch were issued so that students could scurry home to shelter before the storm struck. Over the years I must have heard the statement "the National Severe Storms Forecast Center in Kansas City, Missouri, has issued a tornado watch" hundreds of times. Although I would keep an eye to the sky or turn on the television for information, I never gave much thought to the men who were issuing the forecasts until I read that tornado forecasts were forbidden until 1938. That piqued my interest. I wanted to know why forecasts were forbidden and what caused the change in policy. The result was research that led not only to the story of tornado forecasting but also to its companion, tornado warnings.

This book would have been impossible to write without the assistance of those who issue the tornado watches for the United States, the meteorologists at the Storm Prediction Center in Norman, Oklahoma. Among the many who gave unselfishly of their time and expertise to help me understand the basics of tornado forecasting and provided insights into the history of the center are Joe Schaefer, Robert Johns, and Steve Weiss. Stephen Corfidi graciously showed me the center's daily operations, shared his knowledge of its history, answered endless questions, and provided photographs of current operations. Charlie Crisp supplied much of the material and information on Robert Miller, without which my work would have been much more difficult.

Meteorologists at other locations throughout the country, especially Robert Maddox, Gifford Ely, and José Garcia, pointed me to information long buried in boxes or collecting dust on shelves in various National Weather Service offices. A special thanks belongs to Richard Williams at the Aviation Weather Center in Kansas City, who provided me with many of Joseph Galway's personal papers.

A thank you belongs to Mrs. Beverly Miller, wife of Colonel Robert Miller, for allowing me to use her late husband's unpublished manuscript on the first successful tornado forecasts and to the American Meteorological Society for granting me permission to incorporate into the book my article "Historical Roots of Modern Tornado Forecasts and Warnings," which appeared in *Weather and Forecasting* in 1999.

Many others deserve my gratitude, including Joe Eagleman at the University of Kansas, who suggested I write about tornadoes, Anthony Stranges at Texas A&M University, who continually offered advice, and my longtime friend and colleague Michael Whitley, who provided endless help and encouragement. Among the many librarians and archivists who merit thanks are those at the National Oceanographic and Atmospheric Administration Central Library, the National Archives II in College Park, Maryland, and the National Archives Branch Depository at Kansas City, Missouri. Dodie Guffy, librarian in Texas A&M's Oceanography and Meteorology Department's working collection, located and copied materials and suggested possible sources.

The credit for my passion for weather belongs to my mother,

an avid amateur weather enthusiast who loved to watch the Weather Channel. Finally, I want to express my deepest appreciation to my husband Bill, who for several years patiently endured the trying times and the take-out meals while this work was in progress, and to my children Meredith and Todd, who spent many hours of their childhood huddled in the bathtub when tornadoes were bouncing around.

Scanning the Skies

List of Abbreviations

AFOS	Automation of Field Operations and Services
AMOS	Automatic Meteorological Observing System
AMS	American Meteorological Society
ASOS	Automated Surface Observing System
AWIPS	Advanced Weather Interactive Processing System
AWS	Air Weather Service
ESSA	Environmental Science Services Administration
FAA	Federal Aviation Administration
GOES	Geostationary Operational Environmental Satellite
NEXRAD	Next Generation Radar
NOAA	National Oceanic and Atmospheric Administration
NSSFC	National Severe Storms Forecast Center
NSSL	National Severe Storms Laboratory
NSSP	National Severe Storms Project
NWS	National Weather Service
SELS	Severe Local Storms (Forecast Center)
SPC	Storm Prediction Center
SWU	Severe Weather Unit
SWWC	Severe Weather Warning Center

INTRODUCTION

April and May 1997, peak tornado season in the southern Mississippi Valley and Texas, had been relatively quiet, but when a weak cold front pushed toward central Texas on May 27, the Storm Prediction Center (SPC) in Norman, Oklahoma, issued Tornado Watch 338 for a large area of the state effective from 1:15 P.M. until 7:00 P.M. central daylight time. Television stations in Waco and Austin and radio stations throughout the region forwarded the watch information to their audiences. A large, well-trained, and well-equipped spotter network spread out to find the best vantage point for watching the storms develop. At 1:21 P.M., the National Weather Service (NWS) in Fort Worth issued a tornado warning for McClennan County. Meteorologist Bruce Thomas of television station KCEN in Waco went on the air with live broadcasts of the threatening weather situation and for several hours followed the supercell's progress through McClennan and Bell Counties, where it produced several weak tornadoes that destroyed barns, a marina, and a few homes.

As the storm approached Williamson County, it came within the county warning area of the San Antonio/Austin NWS office. When Doppler radar indicated a tornado about five miles west of Jarrell and spotters on the ground confirmed the storm was moving toward the town of about four hundred just forty miles north of Austin, the NWS office issued a tornado warning at 3:30. The town's warning sirens sounded, and citizens, following the television meteorologists' instructions or recalling what they had learned about tornado safety, took cover under mattresses and blankets in bathrooms or closets. Everything had proceeded according to script, but the tornado bearing down on the town had not read the play. As residents of the Double Creek subdivision on the town's northwest side huddled in their homes, the tornado exploded in intensity and size. The original pencil-thin, snakelike funnel cloud expanded to a black wall half a mile wide. Winds increased to more than 260 miles per hour. Those unfortunate enough to be in this monster's path had little chance of survival. After the twister's passage, townspeople rushed to the scene of devastation and found twenty-seven of their own dead.

For days afterward, television and print journalists asked whether Jarrell had received adequate warning. The obvious answer was "yes." All those involved had done their part to try to protect the people of the tiny central Texas town. A valid watch was in effect. Spotters had observed the tornado and had radioed its position and direction to local law enforcement offices and television stations. The NWS offices in Fort Worth and Austin/San Antonio had tracked the twister on Doppler radar and issued warnings based on radar observation and spotter verification. The local sheriff's department had activated the tornado sirens, and the Waco and Austin television stations had urged people to take shelter. Citizens had heeded the warnings and taken the necessary precautions. Every part of the NWS integrated tornado warning system (forecasting, detection, warning, and response) had gone according to plan.

The only unknown factor had been the tornado itself. NWS meteorologists rated the tornado that struck Jarrell an F5, the highest rating on the Fujita scale. Only the strongest buildings can survive a direct hit by an F5 tornado. By definition these "incredible" tornadoes, packing winds from 261 to 318 miles per hour, will lift strong frame houses off foundations, carrying them considerable distances to disintegrate, and will badly damage steel-reinforced concrete structures. In spite of adequate warning and knowledge of what to do, the people of Jarrell directly in the tornado's path had little chance of surviving because their houses, like most in central Texas, lacked basements or storm cellars. Fortunately, only about 0.2 percent of all tornadoes fall into the F5 category, but the problem lies in determining before they strike which tornadoes will merit "incredible" designation. Although mobile Doppler radar research units have the ability to determine the velocity of winds within a tornado, it is not technologically or economically feasible to forecast each tornado's wind speed, and a tornado's appearance does not divulge its ferocity. The only way to save lives is to be prepared for every tornado.

Each year in the United States, about one thousand tornadoes descend from the clouds, destroying property and lives. Nature's most violent and unpredictable storms have claimed nearly eighteen thousand American lives since 1880. In 1925 the Tri-State Tornado (Missouri, Illinois, and Indiana) alone caused 695 deaths, and several others have taken more than one hundred lives each. As the population increased in the tornado-prone areas, the chances of great loss of life from an individual storm also increased.

To decrease injuries and deaths, the United States instituted a system of tornado watches and warnings to give the citizens of these locales a chance to survive the potentially deadly twisters. An adequate lifesaving system requires more than just shouting "Tornado coming!" SPC meteorologists examine climatic data and atmospheric conditions to determine whether the parameters necessary for tornado formation are present. When they are, the SPC issues a tornado watch (forecast) for a designated area. Immediately volunteer spotter networks and the latest radar, located at NWS offices and many local television stations, begin scanning the skies for signs of the potentially deadly storm. If an individual sees a tornado or radar indicates one might be forming, the warning must go quickly to the public through whatever communication systems are available. Finally, the public, warned of the tornado's approach, must take appropriate action to prevent injury or loss of life.

A successful integrated tornado warning system requires the interaction of the government, scientists, media, educators, and the general public to ensure the effective operation of all four of its components: forecasting, detection, warning, and response. This book relates the story of how the integrated tornado warning system, a unique program that starts at the federal level and extends through state and local governments and the privately controlled broadcast media to the individual family, evolved from its meager beginnings into a program that impacts millions of American lives yearly.

CHAPTER 1

TORNADO THEORIES

An essential component of tornado forecasting is knowledge of the atmospheric conditions that generate this awesome natural phenomenon. Although tornado forecasting is a relatively recent development in human history, the theories of tornadogenesis, or how tornadoes form, occupy a long segment on history's time line. Several ancient cultures developed legends and myths describing how nature's most ferocious storm originated. Scientists from Aristotle to Benjamin Franklin offered theories, but not until the nineteenth and twentieth centuries did serious debate on the causes of tornadoes begin in the United States. Although meteorologists have not yet reached a consensus on how and why certain thunderstorms produce tornadoes, their continued investigations have given tornado forecasting a strong base on which to build.

More than five thousand years ago inhabitants around the Stonehenge area of England, worshipers of the Storm God and Mother Earth, dug more than fifty *cursuses*, or earthwork monuments. Archaeologists and historians have been unable to explain the significance of the cursuses, but British meteorologist Terence Meaden believes that these ditches parallel tornado tracks. To the ancient people of this area, a tornado touching the earth symbolized the mating of the Storm God with Earth Goddess or Mother Earth; to honor this union they dug trenches to serve as monuments to the Storm God's visit to earth.[1]

Several American Indian tribes, especially those on the Great Plains, have legends of the tornado. In a Blackfoot story an orphan boy rescues people whom the tornado Windsucker, depicted as a giant sucker fish, has devoured. The boy brandishes a knife while dancing around inside Windsucker and destroys the monster. In a Seneca legend, the tornado, Dagwa Noenyent, was a giant rolling head that could tear large trees from the earth. Kiowas claimed that they created the tornado and could prevent it from harming them. According to their legend, one summer the weather was so hot and still that the people asked their medicine man for relief. He told the women to bring red mud from the riverbank, and then he shaped the mud into an animal resembling a horse with four legs and a tail. When the medicine man blew into the horse's nostrils, it began to grow, stretch, and twist. The healer named the horse Red Wind and commanded the beast to cool the people. Breaking away from its restraints, the horse whirled through the air, stirring it and blowing up dust. The people, cooled by the wind, became frantic when the gusts blew the feathers off their heads and snapped trees at will. They shouted that the red wind would tear up the earth and blow all of them away. In response to their cries, the medicine man commanded, "Red Wind! I made you. I named you. Now I give you a home. From this time on, you will live in the sky." Ever since its creation, Red Wind, the whirlwind, has lived among the black clouds, but occasionally it twists and stretches down to earth.[2]

Perhaps the first written account of a tornado appears in the Old Testament book of Second Kings when the prophet Elijah "went up by a whirlwind into heaven" (2 Kings 2:11). G. B. Bathurst, in "The Earliest Recorded Tornado?" questions whether Elijah's whirlwind and the pillars of fire and smoke (Exod. 13:21–22) were actually tornadoes, but he does believe that the book of Ezekiel contains a detailed description of a tornado sighting. In verse 4 of the first chapter, Ezekiel writes, "Then I looked, and behold, a whirlwind was coming out of the north, a great cloud with raging fire engulfing itself." Bathurst proposes that verses 5 through 11 depict multiple vortices, although the translators, ignorant of tornadoes, called them "living creatures."[3]

Scientific writings on the theories of tornado development may be as old as the study of meteorology itself. Aristotle wrote in *Meteorologica* that the sun's heating caused the earth to emit two kinds of exhalation, one warm and dry from the earth itself and the other cool and moist from the water on the earth. A predomination of cool and moist exhalation would produce rain; a predomination of warm and dry would produce wind. The collision of opposing winds could create a circular eddy in the clouds, and a whirlwind would appear when "the spiral sinks to the earth and carries with it the cloud from which it is able to free itself. Its blast overturns anything that lies in its path, and its circular motion whirls away and carries off by force anything it meets."[4] Although Aristotle called this spiraling storm a whirlwind, he was apparently describing a tornado.

Pliny the Elder probably was describing a tornado when he wrote in *The Historie of the World*, "if the clift or breach [in the clouds] be not great so that the wind be constrained to turn round, to rol and whirle in discent, without fire, that is to say lightning, it makes a whirlpuffe or ghust called Typhon." The storm, also called Vortex, "snatcheth up whatsoever it meeteth in the way aloft into the skie, carrying it back, and swallowing it upon high." To destroy the approaching storm, Pliny suggested "casting of vinegre out against it as it commeth."[5]

Although tornadoes occurred occasionally in Europe,[6] they were not sufficient in number to cause great concern or interest, but movement to the New World opened the field for exploration. Massachusetts governor John Winthrop, a weather enthusiast who maintained weather logs on the trip from England to the colonies, described in his journal what may be the first recorded tornado in the American colonies. His July 5, 1643, entry notes that a sudden violent wind gust blew down trees and "lifted up their meeting house at Newbury, the people being in it." God's mercy spared all in the building from harm, but one American Indian died when a tree fell on him.[7] Whether or not this was actually a tornado is unknown, but the Reverend Increase Mather in *Illustrious Providences* included an eyewitness account of a July 1680 storm in Cambridge, Massachusetts, that in all probability was the first written record of an actual tornado in the United States. Eyewitness Matthew Bridge declared that a thick black cloud in continuous circular motion produced a great noise in the process of tearing down trees and picking up bushes, trees, and large stones. Inside the cloud "there seemed to be a light pillar as he judged about eight or ten feet in diameter, which seemed to him like a screw or solid body." John Robbins, a servant, died in the storm.[8] Tornadoes, however, were not common occurrences in the American colonies. David Ludlum lists only twenty pre–Revolutionary War tornadoes in *Early American Tornadoes, 1586–1870*. Massachusetts recorded eight of the storms before 1776, while several other colonies reported one or two tornadoes. Only four storms produced deaths, with a South Carolina tornado in 1761 killing eight.[9]

This rare phenomenon generated only limited interest among scientists during the colonial period. Articles in *Transactions of the American Philosophical Society* were the main source of colonial American writings on tornadoes. John Perkins, a Boston physician interested in winds, corresponded with Benjamin Franklin in 1752 and 1753 on the subject of waterspouts. Perkins believed that all violent storms resulted from cold air descending from aloft and that tornadoes and hurricanes were the same type of storms differing in specifics such as appearance. Dissatisfied with the word *tornado*, which primarily referred to violent storms at sea, Perkins proposed that the violent windstorms that had been observed a few times in the northern colonies should have a distinct name, such as *wind-spout*. He described this violent act of nature as a sudden storm defined by a "spout of wind coming from it that strikes the ground in a round spot of a few rods or perches in diameter, with a prone direction, in the course of the wind of the day, and proceeds thus half a mile or a mile."[10]

Benjamin Franklin, American scientist and statesman, made one of his most striking meteorological observations in 1743: Storms in the United States generally move from west to east or southwest to northeast. Franklin's letter to Alexander Small on May 2, 1760, explained that he had developed the hypothesis about twenty years earlier, when thick storm clouds over Philadelphia had hindered his attempt to observe a lunar eclipse. Franklin had assumed that the storm began earlier in the area northeast of the Pennsylvania city, but after his brother in Boston informed him that the storm moved into that area after the eclipse, he checked reports from the other colonies and noticed that the storm had impacted Philadelphia before it affected New York or other points northeastward. Franklin concluded that he "found the beginning to be always later the farther northeast" and that "to produce our northeast storms, I suppose some great heat and rarefaction of the air in or about the Gulf of Mexico." His interest in meteorology led to theories on rain, lightning, waterspouts, and whirlwinds. Franklin believed a thunderstorm required three elements: particles of air, particles of water, and electricity. The accepted theory of Franklin's day was that electricity, like heat, light, and magnetism, was a weightless fluid that all bodies contained. When a body with a deficit of electric fluid, or electric fire, came into contact with a body with excess electric fire, the fluid would flow, like water, from the body with the excess to the body with the deficit, generating a spark. Water particles rising from the ocean contained electric fire. Should a cloud loaded with this electric fluid (fire) encounter a mountain or another cloud system lacking electric fire, a "snap of lightning" would release the electricity and restore equilibrium.[11]

Whether Franklin ever saw a tornado is unknown, but he believed that waterspouts and land whirlwinds were the same.

In a February 4, 1753, letter to John Perkins, Franklin explained that whirls, whether ascending or descending, were of the same origin. He believed that the heavier cold air in the upper regions pushed down on and displaced the lighter warm air near the earth's surface. Heavy fluids descending, like water down a drain, frequently formed whirls or eddies. The fluid acquired a circular motion, and centrifugal force created a vacancy in the middle "like a speaking trumpet, its big end upwards." Both ascending and descending air formed the same type of eddy, thus causing the appearance of the whirlwind's funnel. The whirling air, receding from the center, created a vacuum that could be filled only from the lower end, not the sides; therefore, the air rushing into the funnel at the lower end carried dust, leaves, and other items in its path.[12]

Franklin and Perkins, like most eighteenth-century scientists, were interested in a broad range of scientific ideas. Science was not specialized; many scientists had training in medicine or natural philosophy, which encompassed a broad range of subjects, from geology to botany. Meteorology was a companion to another observational science, astronomy. With few exceptions, those interested in the weather were simply observers who recorded local weather data, such as temperature and sky conditions.

Although Franklin never directly linked lightning with tornadoes, fellow Pennsylvanian Robert Hare did. In the *American Journal of Science* the University of Pennsylvania chemistry professor expressed his belief that "a tornado is the effect of an electrified current of air, superseding the more usual means of discharge between the earth and the cloud in those sparks or flashes which are called lightning." Hare argued that a tornado was an outgrowth of a thunderstorm. In the typical thunderstorm, a lightning bolt manifested the exchange of electricity between a cloud and the ground, but in a tornado the funnel cloud facilitated the electrical discharge. Hare wrote, "everything proves that the tornado is nothing else than a conductor formed of the clouds, which serves as a passage for a continual discharge of electricity from those above." He went on to explain that an electrical discharge created a vacuum that enabled the surrounding atmospheric particles to rush inward and upward. Tornado damage occurred when rising air currents and the electrical attraction between the earth and the cloud lifted objects from the earth. Incredibly, Hare ignored the obvious, that a tornado's winds caused the damage. Instead, he proposed that just as an electrical attraction in a charged jar produced an up-and-down motion of pith balls or other light objects, lightning, a much greater current, lifted houses, trees, and other objects up and down.[13] Hare's contemporaries gave only limited assent to his electrical theory.

Interest in tornadoes expanded when a twister that struck New Brunswick, New Jersey, on June 19, 1835, generated a dispute between classical language professor James Pollard Espy and transportation engineer William Redfield. Their ongoing debates on the origins of storms and the characteristics of tornadoes appeared in scientific journals for several years. The basic questions were not only what caused tornado formation but also whether tornadoes had a whirling character. Redfield found it remarkable that "previous to this period the evidences of the rotation, or other characteristic action, of tornadoes appear not to have been duly examined and recorded, nor even to have received the distinct consideration of scientific observers."[14]

Although trained in the classics and law, James Espy, the instigator of the debate over the tornado's nature, would become America's leading meteorologist in the mid-nineteenth century. Espy served as professor of languages and head of the classics department at the Franklin Institute, and many of his early writings appeared in the *Journal of the Franklin Institute.* An avid interest in astronomy and meteorology led Espy in 1837 to launch a campaign to establish a network of weather observers throughout the country. Eleven years later, the Smithsonian Institution would initiate such a system. From 1842 to 1857 Espy served as a government meteorologist under the Navy Department's jurisdiction. In addition to his data collection and storm theory, Espy, known as America's Storm King, is best remembered as an advocate of artificial precipitation or "rainmaking."[15]

The core of Espy's meteorological theories was the behavior of air expanded by heat. An interest in dew points and barometric pressure led to his storm theory, which was first published in the 1835 *Transactions of the Geological Society of Pennsylvania.* Using an air pump as his example, Espy stated that an ascending column of air would expand and cool, causing the water vapor present to condense into clouds and rain. Additional heat that the condensation released would expand the remaining air, pushing it higher into the atmosphere, where more vapor would condense as the air expanded. He concluded that "all the phenomena of rains, hail, snows, and water spouts, change of winds and depressions of the barometer follow as easy and natural corollaries from the theory here advanced, that *there is an expansion of the air containing transparent vapour when that vapour is condensed into water.*"[16]

The debate over the whirling characteristic of a tornado began when Espy and Alexander Bache, an avid weather observer and Benjamin Franklin's great-grandson, visited the site of the New Brunswick tornado for five days, interviewing witnesses and taking copious notes on the position and condition of debris. Espy never put his findings in a formal paper, but in a July 17, 1835, address to the American Philosophical Society, he said, "the effects all indicate a moving volume of rarefied air, without any whirling motion at or near the surface of the earth." Bache, agreeing with Espy, wrote that a study of the fallen trees in the tornado's path led him to conclude that "these effects all indicate a moving column of rarefied air, without any whirling motion at or near the surface of the ground."[17] Robert Hare also supported Espy. He did not believe that gyration was an essential feature of tornadoes but conceded that the swirling motion could be a "casual effect of the currents rushing toward the axis of the tornado."[18]

William Redfield's chief interest was hurricanes. The self-educated Connecticut engineer and later first president of the American Association for the Advancement of Science presented his theory on the nature of the North Atlantic coastal storms in the July 1831 issue of the *American Journal of Science*. Redfield declared that a hurricane was a great whirlwind or a rotary windstorm related to tornadoes and waterspouts and dismissed the common theories of storm causation, which involved electricity and rarefied or heated air. He conceived of a storm system as a rotating flat disk that originated when a layer of cold air moved over a layer of warm air. The cold air would descend and "an immediate gyration or convulsion will take place in the two masses at this point." The warm air would rise, a vortex would form, and the cyclone would begin forward motion. In Redfield's model the winds moved around a center in a generally circular path.[19]

Redfield surveyed the New Brunswick tornado scene three times and presented his preliminary findings in the January 1839 *American Journal of Science*. He maintained that his investigation showed "the whirling character of this tornado, as well as the inward tendency of the vortex at the surface of the ground; and further that the direction of this rotation was towards the left, as in the North Atlantic hurricanes."[20] In other words, tornadoes were counterclockwise, whirling windstorms. The simultaneous appearance of Redfield's definitive article on the tornado's whirling nature in the July 1841 issues of the *American Journal of Science* and *Journal of the Franklin Institute* attests to its importance to the scientific community. Through use of maps and sketches of fallen trees Redfield presented his case that a tornado "had the common properties which may be observed in all narrow and violent vortices, viz: *a spirally descending and involuted motion*."[21] Hare and Espy attacked Redfield, who defended his position in numerous articles throughout 1842 before returning to hurricane studies.

William Ferrel, a physical scientist who gave meteorology a foundation in mathematics and fluid dynamics, was interested in atmospheric circulation. His article in the 1861 *Journal of American Science* discussed the motions of the atmosphere resulting from local disturbances such as tornadoes and water-spouts. Ferrel proposed that tornadoes occurred only when the surface of the earth was very warm and the atmosphere very calm. The heated air near the surface was unstable, and when some slight disturbance upset the equilibrium, the warm air would burst up through the lower layers of the atmosphere, "somewhat as the vapor of boiling water, which is generated mostly at the bottom of the containing vessel, bursts up through the water above and comes to the surface." Air would flow rapidly from all sides into the rising air column, and "unless the sum of all the initial moments of gyration around the centre is exactly 0, which can rarely ever be the case, it must run into rapid gyrations near the centre, and a tornado is the consequence."[22]

Most tornado theories had scientific merit. One exception was Theodore Wiseman's theory, which he presented in the short treatise *Origin and Laws Governing Tornadoes, Cyclones, Thunderstorms, and Kansas Twisters*. He proposed that the sun produced only positive electricity, which the negatively charged planets absorbed. In like manner, the sun absorbed the negative electrical charges the planets emitted. Wiseman presented no explanation for his assumptions that the sun and planets produced opposite electrical charges, but he believed the fusion of the positive and negative charges created violent disturbances in the vicinity. During the morning hours the sun would transfer large quantities of positive charges to the upper atmosphere's fleecy clouds. By late afternoon the clouds, heavily laden with positive electricity, sought a negatively charged body. In the absence of negative moist clouds in the atmosphere, the positively charged clouds would be attracted to the negatively charged earth. The rapidly descending positive charges would "attain a rapid rotary motion in their descent, so that these positive currents get a hollow funnel shape, or the shape of an hour glass." Wiseman believed that the electrically charged twenty- to fifty-foot thick circular outer wall of the funnel tore things apart while the inner circle (hollow part of the funnel) carried the debris upward, sometimes thousands of feet. The tornado dissipated when it discharged its superabundance of electricity over a distance of forty or fifty miles or when it encountered a negatively charged cloud.

Essential to Wiseman's theory was his belief that tornadoes could originate and exist only in arid regions, where no negatively charged moisture-laden clouds would attract the positive charge. To validate his ideas, he contended that no tornadoes occurred in Kansas in 1883 or 1884, years in which substantial spring rains fell, whereas the dry years of 1880 through 1882 experienced several tornadoes.[23] An examination of this theory begs an answer to the question of why tornadoes do not appear daily in dry regions, but no rebuttal of Wiseman's ideas appeared in the scientific literature of the day. Perhaps very few read his work, or maybe the theory was just too flawed for anyone to take it seriously.

The American Meteorological Journal, "recognizing the high importance of a fuller knowledge of Tornadoes, and believing that a combined effort will much advance our knowledge," announced a tornado essay competition in its July 1888 issue and offered a two-hundred-dollar prize for the best original unpublished work. The journal suggested that the essays might discuss specific tornado occurrences, tornado prediction, or tornado theory, including the development of knowledge in these areas. Judges would give special attention to the article's scientific value and originality. The winning essays appeared in the August 1890 issue of the journal. The U.S. Army Signal Corps' tornado expert, John Park Finley, captured first place; Alexander McAdie, a Clark University doctoral student finished second; and Henry A. Hazen, Finley's most persistent critic, came in third. Finley, who had already written *Tornadoes* (1887), a book that included extensive statistics as well as rules for observing and surviving tornadoes, offered no theory on tornado formation but explained in great detail the conditions necessary for the storm's development—an unstable atmosphere and a gyratory motion of the air around some center of circulation. He concluded that tornadoes formed in the southeast quadrant of a low-pressure area, where the circulation of warm, moist air underneath the cold air gave rise to unstable atmospheric conditions, but he offered no explanation for why only a few thunderstorms produced twisters.[24] McAdie concurred with Finley's two conditions critical for tornado formation but borrowed a third factor from Ferrel—a local disturbance.[25]

The Signal Corps had hired Hazen as a civilian forecaster in 1881. When Finley fell from favor in 1889, Hazen replaced him as the corps's tornado spokesman. Hazen's book *The Tornado* and the tornado essay prize confirmed his stature as the new tornado authority. In his essay, Hazen refuted the idea that the planets' position relative to the earth and sun caused tornadoes, but he did concede that sunspots might produce a minor increase in tornado activity. In like manner, he dismissed all electrical theories and the widely accepted idea that tornadoes occurred when different air currents of different temperatures clashed. Hazen concluded that "most of the generally accepted theories certainly seem to be controverted by well known facts and also seem entirely inadequate to account for the phenomena observed," but he offered no plausible alternative for tornadogenesis.[26] Although nineteenth-century scientists had identified surface weather features frequently present at a tornado's birth, they had collected no data on the atmosphere's vertical structure around thunderstorms, so at the close of the century meteorologists were still unable to explain the mechanism that triggered twisters.

In 1951, more than sixty years after the publication of Hazen's essay, Edward Brooks, a meteorologist at Saint Louis University, noted that "a major problem in explaining the formation of a tornado is to find the source of the potential energy and the manner in which it is converted into kinetic energy."[27] In the twentieth century numerous new theories (including some rather odd ones) joined electricity and convection, the two major theories of the preceding century.

A ban on the use of the word *tornado* in forecasts, which was in effect from 1885 to 1938,[28] may have been a deterrent to tornado research in the United States. During this period only scattered writings on tornadoes appeared in the literature. Alfred A. Henry and three other Weather Bureau forecasters included only three pages on tornadoes in their 1916 book, *Weather Forecasting in the United States*. They proposed that the tornado's origin was mechanical and its rotation derived from the cyclone (low-pressure area) in which it occurred. Rotation began when adjacent winds from opposite directions intermingled at a kilometer or so above the earth's surface. Occasionally, just in front of a thunderstorm these countercurrents would come close enough together to produce a violent whirl that a strong updraft would sustain. Strangely, the entire three pages appeared verbatim in an April 1920 *Monthly Weather Review* article entitled "The Tornado and Its Cause." This time the listed author was William J. Humphreys, the Weather Bureau's "meteorological physicist," who was not listed among the authors of the 1916 book. This duplication exemplifies the dearth of research and writing on tornadoes during this period.[29]

Humphreys' own work appeared six years later in the *Monthly Weather Review*. He noted that many had described the storm's appearance and effects but that none had provided a satisfactory account of its origin. Humphreys hoped that his inclusion of a list of twenty-six meteorological conditions connected with a tornado's appearance might inspire someone to find the problem's ultimate solution. All of the significant factors, including geographical and meteorological location, types of clouds, precipitation, and lightning, were ground-based, but

Humphreys did include a reference to data the Weather Bureau had collected using sounding balloons and kites in the lower one or two kilometers of the atmosphere.

Humphreys believed that although all observations were too distant from the tornadoes to have much specific value, the kite and balloon readings did provide a clue to the wind velocity and direction in the atmospheric layer where the severe storms appeared to originate. Humphreys reasoned that two key ingredients in tornado formation were changes in wind direction with height (wind shear) and instability at one or more levels in the atmosphere; in other words, "there are adjacent, presumably superjacent, currents of air of different sources where and whenever tornadoes are likely to occur." Tornadoes most frequently developed when these air currents were a humid southerly wind and an approaching midlevel cold front with northwesterly winds. Humphreys' willingness to include the relatively new concept of a cold front in this article is strange, considering that his widely used textbook, *Physics of the Air*, devoted less than one page to airmass and frontal analysis.[30]

Weather fronts were a relatively new concept. During World War I, the warring nations banned the dissemination of weather data and forecasts. Neutral Norwegians, cut off from all essential information, sought new methods of weather forecasting. Vilhelm Bjerknes, a physicist who applied hydrodynamic and thermodynamic theories to meteorology, headed meteorological research at the Bergen Museum's Geophysics Institute. Bjerknes and his son Jacob, realizing the importance of weather forecasting to Norway's food production, studied summer cyclones, or low-pressure systems, and initiated weather forecasts for Norway's farmers. Father and son explored the fact that most weather disturbances occur at the boundaries between air currents of different humidity and temperature and adopted the term "front" to signify a type of battle between opposing air masses. Bjerknes realized that almost every kind of weather change resulted from the passage of a front, whether warm or cold, and that knowledge of these phenomena would be essential for practical weather forecasting, but the U.S. Weather Bureau did not officially adopt his theory until 1936, when the familiar fronts began to appear on weather maps. The bureau may have been slow to adopt the Bergen institute's new methods, models, and theories in the early 1930s because two of its three meteorologists with doctorates, including Humphreys, were over seventy years of age and preferred to use their traditional cartographic and analytical methods to prepare the daily forecasts.[31]

Humphreys' connection between a cold front and tornado formation became the dominant theory through the 1940s. J. R. Lloyd, Kansas City Weather Bureau station chief, validated Humphreys' concept in his 1942 *Monthly Weather Review* article. Lloyd concluded that "tornadoes appear to occur only in connection with upper-air cold fronts" and "remain on the upper-air cold front throughout their existence." To confirm his idea Lloyd attributed the seasonal migration of tornadoes northward to the disappearance of the polar cold front in the south.[32]

Morris Tepper, Weather Bureau Scientific Services division chief in Washington, D.C., proposed in 1950 that tornadoes formed at the intersection of two unequal pressure jump lines. He defined a pressure jump as "a sudden rise in barometric pressure as recorded on the trace of a barograph." When these two lines interacted, a sharp wind shear occurred and generated a vortex that could develop into a tornado if suitable thermodynamic conditions were present. In Tepper's opinion, Finley's observations that tornadoes formed when clouds from the northwest and southwest met supported his theory. To test Tepper's hypothesis the Weather Bureau operated the Tornado Project in 1951 and 1952 to record barometric pressure readings at 135 sites in Kansas and Oklahoma. Thirty complete weather stations within the designated area took other atmospheric measurements, the U.S. Air Force provided radar coverage, and weather observers reported all tornado occurrences. During these two years 27 percent of the 143 tornadoes occurred along pressure jump lines, but not necessarily at the lines' point of intersection. Tepper abandoned his theory in 1954.[33]

In 1953 the United States recorded 421 tornadoes, nearly twice the number of any previous year. Many Americans wondered whether nuclear explosions produced the increase because more than half of the year's tornadoes, including the catastrophic storms in Waco, Texas; Flint, Michigan; and Worcester, Massachusetts, formed between March 17 and June 15, the period of atomic weapons testing in Nevada. D. Lee Harris of the Washington, D. C., Weather Bureau Office studied data on atomic explosions and tornadoes between 1951 and 1954 and concluded that the statistics did not "indicate a tendency for a relative increase in tornadoes during periods of atomic explosions."[34]

Since World War II, numerous meteorologists have filled in pieces of the tornadogenesis puzzle. Edward Brooks observed several tornadoes that passed near Saint Louis during the spring of 1948. Of special interest were the barometric readings and wind speed recorded at the airport two and one-half miles southeast of the March 19, 1948, tornado's path. Within a fifteen-minute period the barometric pressure dropped three millibars and then suddenly rose five millibars. The wind speed was seventeen miles per hour with an occasional gust of twenty-two miles per hour from the southeast. At greater distances from the tornado, winds gusted at over fifty miles per hour. From these observations Brooks proposed that the actual funnel cloud, much like the eye of a hurricane, existed within a much larger circulation pattern he designated a "tornado cyclone" (now called a *mesocyclone*). A tornado could form within this air mass "intermediate in size between the parent low and the tornado funnel itself" if the rate of airflow into the system exceeded the loss to friction.[35]

While visiting the National Severe Storms Project in the early 1960s, British meteorologist Keith A. Browning found another piece of the puzzle. From a study of radar data he realized that most tornadoes form inside very large and vicious storms he called *supercells*. Within this cloud structure of one thousand cubic miles or greater, which often towered fifty to sixty thousand feet into the atmosphere, was a special three-dimensional airflow pattern that allowed the storms to have long lives and reach massive sizes.[36] If winds in the higher layers of the atmosphere blew stronger than those in the lower or boundary layer, the air between the layers was rolled much like a rolling pin. When this horizontal rolling tube of air encountered the supercell's updraft, the "rolling pin" tilted upward and spun on its end. The entire updraft would then begin to spin and give birth to the mesocyclone. Many mesocyclones spawn tornadoes, and these are frequently the strong and violent type that most often cause death and massive destruction.[37]

Occasionally, tornado witnesses reported more than one funnel present in the same storm. Two researchers in the early 1970s explored the formation of multiple vortices, or smaller whirlwinds, that rotated around the main funnel. Neil Ward, researcher at the National Severe Storms Laboratory (NSSL) in Norman, Oklahoma, perfected a laboratory tornado model that could duplicate the change of a single vortex tornado into a multiple vortex one. From a study of aerial photographs of ground markings made by tornadoes, Theodore Fujita[38] at the University of Chicago concluded that multiple vortices, or what he termed *suction vortices*, existed. These suction vortices, sometimes called *suction spots*, explained why one house might be completely destroyed while the one next door was virtually untouched. The fortunate house was in the area between the vortices.[39]

In addition to his theoretical work, in 1970 Fujita devised a tornado rating scale that used structural damage to estimate wind speeds within tornadoes (table 1). Since 1971 National Weather Service (NWS) meteorologists have assessed every reported tornado and assigned a Fujita or F-scale rating based

TABLE 1

The Fujita Scale

F-Scale Number	Intensity Phrase	Wind Speed	Type of Damage Done
F0	Gale tornado	40–72 mph	Some damage to chimneys; breaks branches off trees; pushes over shallow-rooted trees; damages sign boards
F1	Moderate tornado	73–112 mph	The lower limit is the beginning of hurricane wind speed; peels surface off roofs; mobile homes pushed off foundations or overturned; moving autos pushed off the roads; attached garages may be destroyed
F2	Significant tornado	113–157 mph	Roofs torn off frame homes; mobile homes demolished; boxcars pushed over; large trees snapped or uprooted; light object missiles generated
F3	Severe tornado	158–206 mph	Roof and some walls torn off well-constructed houses; trains overturned; most trees in forest uprooted

TABLE 1 (con't)

The Fujita Scale

F-Scale Number	Intensity Phrase	Wind Speed	Type of Damage Done
F4	Devastating tornado	207–260 mph	Well-constructed houses leveled; structures with weak foundations blown off some distance; cars thrown and large missiles generated
F5	Incredible tornado	261–318 mph	Strong frame houses lifted off foundations and carried considerable distances to disintegrate; automobile-sized missiles fly through the air in excess of 100 meters; trees debarked; steel-reinforced concrete structures badly damaged

Source: Thomas Grazulis, *Significant Tornadoes, 1680–1991*, 141.

on the single most intense example of damage in its path. In addition, the National Severe Storms Forecast Center (NSSFC) in Kansas City, Missouri, used photographs and damage reports to retroactively assign F-scale ratings to all post-1950 tornadoes in its database. When the NSSFC initially accepted the use of the Fujita scale, its director, Allen Pearson, added scales for path length and path width (table 2), which gave every tornado a three-number designation or F, P, P rating (Fujita intensity scale, Pearson path length scale, and Pearson width scale).[40] In practice, the Pearson scales were not as widely employed as the Fujita scale, and they soon disappeared from common use.

One of the more unusual theories of tornado development appeared in a 1975 *Nature* article. John D. Isaacs, James W. Stork, David B. Goldstein, and Gerald L. Wick, researchers at the University of California's Foundation for Ocean Research at San Diego, proposed that the sixfold increase in tornado occurrences in the United States during the preceding four decades corresponded to a similar increase in motor vehicle traffic. They believed that the counterclockwise torque generated by streams of cars and trucks passing each other on the right created a cyclonic vorticity in the atmosphere. According to the study, fewer tornadoes occurred on Saturday than on any other day of the week because most truck and commuter traffic ended on Friday, and Saturday travel patterns were more random than those during the week, when many vehicles poured into urban areas. The scientists concluded that if their findings were true, "at least 14% of US tornadoes are under man's control," but they did not have enough faith in their findings to propose that Americans begin driving on the left.[41] In jest Allen Pearson responded to Isaacs's theory with his own suggestion: Perhaps the popularity of X-rated movies that overheated the air brought tornadoes because they needed hot, steamy air to form.[42]

The Warner Brothers 1996 movie *Twister* brought the story of tornado chasers to the screen. Some chasers hunt tornadoes merely for the thrill, but many are scientists who are trying to

TABLE 2

The Fujita-Pearson FPP Scale

SCALE	F (MPH)	P LENGTH (MILES)	P WIDTH
negative	less than 40	less than 0.3	less than 6 yards
0	40–72	0.3–1.0	6–17 yards
1	73–112	1.0–3.1	18–55 yards
2	113–157	3.2–9.9	56–175 yards
3	158–206	10–31	176–556 yards
4	207–260	32–99	0.3–0.9 miles
5	261–318	100–315	1.0–3.1 miles

SOURCE: Theodore Fujita and Allen D. Pearson, "Results of FPP Classification of 1971 and 1972 Tornadoes," in *Preprints of Eighth Conference on Severe Local Storms,* 142.

take measurements of various meteorological phenomena in the storms in an attempt not only to understand their actions better but also to understand why they form. Although the majority of tornadoes are small, weak, short-lived storms that frequently develop in conjunction with a squall line or a tropical storm, much of the meteorologists' attention has focused on the violent supercell tornado, the often long-lived, strong harbinger of death and destruction. Doppler radar and computer modeling have aided in the study of these monsters, but progress in understanding the details of their formation has been slow because the storms are not as common as their weak brothers and because researchers face danger when they try to obtain direct measurements from the storm.

To aid in obtaining data from these storms, intercept teams from the University of Oklahoma, led by Howard Bluestein, attempted to place a four-hundred-pound instrument package—designated TOTO (Totable Tornado Observatory) in honor of Dorothy's dog in Frank L. Baum's *The Wizard of Oz*—directly in a tornado's path. Al Bedard of the Environmental Research Laboratory in Boulder, Colorado, designed TOTO, which is about the size of a fifty-five-gallon oil drum, so that the team could mount it in a pickup truck, transport it to the location of a developing tornado, and quickly set it up. The package, containing instruments to record wind speed, wind direction, temperature, and static pressure, supposedly would stay put even if a tornado struck it directly. Although Bluestein and his crew tried from 1981 through 1983 to place TOTO in the direct path of a tornado, they failed. So did teams from the NSSL during 1984 and 1985.[43]

Computers added another dimension to the exploration of tornadogenesis: computer modeling or simulations. Robert Williamson of the University of Illinois and Joseph Klemp of the National Center for Atmospheric Research (NCAR), through the use of numerical solutions to various equations, simulated a three-dimensional supercell on a computer in 1978. Their work and later computer modeling "debunked the popular explanation that tornadoes are caused by colliding air masses" and showed that the earth's rotation had little effect during a storm's initial stages. Modeling did establish that wind shear, or change of wind direction or speed with height, was crucial to the development of rotation. Klemp and Richard Rotunno of NCAR used computer modeling in 1985 to demonstrate that low-level rotation in a supercell depended on its cooled downdraft. Similarly, Robert Davies-Jones and Harold E. Brooks at NSSL demonstrated in 1993 that this cooled air "flows along the surface and is sucked up into the southwest side of the updraft. Because the flow to the updraft is convergent, the air rotates faster, like an ice skater who spins more rapidly by drawing in her arms." Although these computer simulations revealed possible mechanisms for large-scale rotation at middle and low levels of a mesocyclone, they did not offer a definitive reason for tornado formation.[44]

Meteorologists understand the process of supercell and mesocyclone formation, but they do not agree on the mechanism that produces the tornado itself. Although most acknowledge that an updraft is essential, many doubt that it is the sole factor in tornado formation. One group thinks downdrafts in the rear of the storm provide the key, whereas another faction believes that tornadoes, like their weaker relatives, the waterspout and dust devil, build their vortices from the ground upward. To determine the correctness of these theories the NSSL and the Center for Analysis and Prediction of Storms conducted a large-scale observation program during the springs of 1994 and 1995. According to field commander Eric Rasmussen, the NOAA/National Science Foundation-funded project VORTEX (Verification of the Origins of Rotation in Tornadoes Experiment) "tried to gather data about one of the last great mysteries of tornado formation, that final step that takes the vortex from high altitude rotation inside the supercell thunderstorm to a tornado on the ground." During the two-year project, meteorologists from many major universities and weather research facilities in the United States and Canada employed an armada of vehicles, including specially equipped vans and a P-3 hurricane hunter airplane, to chase supercells throughout the Texas Panhandle, Oklahoma, and Kansas. In June 1995 the team employed mobile mesonets (vehicles equipped to measure a variety of meteorological phenomena), airborne radar, multiple video cameras, and a mobile Doppler radar (Doppler on Wheels) to accumulate an abundance of data when tornadoes swarmed across the Texas Panhandle. The project left meteorologists with so much data that Rasmussen estimated "it might take five to ten years to search all of the raw data and locate all the clues that will lead to new ideas about tornado formation."[45]

Rasmussen published in the summer/fall 1998 *NSSL Briefings* a few important new findings he, David Blanchard of NSSL, and Jerry Straka and Paul Markowski of the University of Oklahoma gleaned from the VORTEX data. They found that "boundaries, which are the leading edges of pools of cooler air left behind by thunderstorms, are prime locations for later tornado formation" and that "tornado formation itself seems to be strongly linked to the character and behavior of a downdraft at the back side of the supercell storm." In addition, VORTEX demonstrated that although scientists had believed that as many as one-half of all mesocyclones produced tornadoes, this was not the case. Rather, "tornado formation is a complicated process that depends perhaps only indirectly on the presence of a mesocyclone" and "the difference between tornadic and non-tornadic mesocyclones can be very, very subtle."[46]

Joshua Wurman of the University of Oklahoma was the project leader for the Radar Observations of Tornadoes and Thunderstorm Experiment (ROTATE) in 1998, a program that employed two Doppler on Wheels mobile radars to observe tornadoes as closely as possible to study their formation, structure, life cycle, and death. The dual mobile Dopplers permitted the mapping of tornadic winds and the observation of detailed structures on a much smaller scale than other Doppler radars allow. The main focus was to collect additional data that might aid in evaluating the numerous tornadogenesis theories that projects such as VORTEX advanced.[47]

During the last two centuries meteorologists have attempted to determine what causes a tornado to form. Prominent theories have included electricity, convection, and pressure change. More esoteric offerings, such as atomic explosions and automobile traffic, were short lived. Perhaps an examination of the data from the VORTEX project along with computer modeling will enable future researchers to determine the steps in tornado formation that in turn will provide an invaluable tool for tornado forecasters.

CHAPTER 2

TORNADO FORECASTING TO 1940

Although Benjamin Franklin realized as early as 1743 that storms in the United States generally moved from west to east or southwest to northeast, the lack of rapid communications hindered warning regions in a storm's path. It was not until a century later, when the first commercial telegraph line was opened on April 1, 1845, that the possibility of forewarning communities of approaching severe storms and tornadoes was raised. Joseph Henry, America's premier physicist and the first director of the Smithsonian Institution, proposed in a letter to the regents dated December 8, 1847, that the institution "organize a system of observation which shall extend as far as possible over the North American continent." In the *Smithsonian Report* of the same year he wrote that the time was "auspicious for commencing an enterprise . . . an extensive system of meteorological observations, particularly with reference to the phenomena of American storms." The following year Henry initiated a volunteer weather observation program. He used the one-thousand-dollar budget allowance from the Smithsonian regents to purchase weather instruments, chiefly thermometers and barometers, and judiciously distributed them to observers throughout the country. By 1852 the Smithsonian had more than two hundred such volunteers. Henry, whose work in electromagnetism was the foundation for telegraphy, persuaded the telegraph companies to transmit, at no cost, weather reports from volunteer observers, military posts, and surveying parties throughout the nation to the Smithsonian in Washington, D.C., where he synthesized the reports into a daily weather map. Henry began displaying weather signals on the building's high tower and posting a large weather map for the public on a wall of the Smithsonian in 1850. The *Washington Evening Star* began issuing newspaper weather reports on May 7, 1857, but these reports were all after the fact. The Smithsonian made no attempt to forecast the weather.[1]

By 1860 the Smithsonian volunteer network encompassed more than five hundred reporting stations. A series of devastating spring and summer tornadoes that year prompted Henry to request information from eyewitnesses. In 1862 the Smithsonian distributed circulars to the public warning of the dangers posed by tornadoes and asking for continued reports and data on these storms. The public's response was so great that in 1872 the institution issued a four-page pamphlet listing the questions observers should attempt to answer when reporting a tornado, including the date, location, length and width of path, direction and speed of movement, color of sky, and shape of funnel.[2]

In spite of the Smithsonian's admirable effort to collect storm data, no one used the information to produce weather forecasts. Cleveland Abbe, director of the Cincinnati Astronomical Observatory, suggested issuing daily weather forecasts and storm warnings. In his address to the observatory staff on May 1, 1868, Abbe said that with "a proper system of weather reports the public need of forecasts could be met and that astronomy also could be benefitted." The director hoped that by providing a useful service to the community the observatory could improve its ranking among the world's observatories. He called for the observatory to "keep a record of regular hourly observations of all phenomena depending upon the atmosphere." Concerning weather forecasting, he concluded that "although we cannot yet predict the weather for a week in advance, we are safe in saying that with a proper arrangement of outposts, we can generally predict three days in advance any extended storm and six hours in advance any violent hurricane." He proposed collecting daily telegraphic reports of weather conditions from a network of stations much as the Smithsonian had done and incorporating this data into a "prediction" that newspapers could disperse to the public. Using telegraph reports from only Saint Louis and Leavenworth, Kansas, Abbe published the first forecast in the September 2, 1869, daily *Weather Bulletin*: "Easterly and Southerly winds prevail. Barometer has begun to fall at Cincinnati and a storm passing over the southern country will not reach Cincinnati. Clouds and warm weather this evening. Tomorrow clear." With these few words weather forecasting in the United States began. Abbe continued his experiment for three months.[3]

In 1868 storms on the Great Lakes destroyed or damaged 1,164 vessels and killed 321; the following year 209 perished. Increase A. Lapham of Milwaukee, a veteran Smithsonian weather observer and Abbe supporter, believed that a chief cause of loss of life was the lack of forewarning. He asked Milwaukee congressman Halbert E. Paine whether it was not the duty of the government to try to prevent such loss of life and property. In response, Paine introduced a bill in December 1869 authorizing the secretary of war to provide for taking meteorological observations at all military stations and for notifying shipping interests along the Great Lakes and Atlantic Coast of approaching storms. Some scientists, particularly Abbe, had hoped that a weather service would be in the hands of civilians rather than the army, but Paine argued that the military could provide the service at lower cost. While the debate about civilian versus military control continued, Paine introduced a joint resolution containing the provisions of the original bill, which the House of Representatives passed on February 2, 1870; the Senate concurred three days later. On February 9, President Ulysses S. Grant signed the joint resolution that authorized and required the secretary of war "to provide for taking meteorological observations, at the military stations in the interior of the continent, and at other points in the States and Territories of the United States, and for giving notice, on the northern lakes and on the seacoast, by magnetic telegraph and marine signals, of the approach and force of storm."[4] These few words gave birth to a national weather service that, through its forecasts and warnings, would touch the life of every American.

The responsibility for creating the new weather service lay with Brigadier General Albert J. Myer of the Army Signal Corps, who named the service the Division of Telegrams and Reports for the Benefit of Commerce.[5] Stations throughout the country sent three atmospheric condition reports daily to the telegraph room of its Washington, D.C., headquarters. Some metropolitan newspapers, including Washington's *Evening Star*, began inserting several reporting posts' morning observations on their front pages beginning November 1, 1870. The initial twenty-four field stations telegraphed only the state of the weather (fair or cloudy), but as the number of stations increased to 284 by 1878, the reports expanded to include barometric pressure, temperature, humidity, wind velocity and direction, amount of cloud cover, and general weather conditions.[6]

Although military personnel made the observations, weather predictions and storm warnings were in the hands of civilians the corps had hired. Increase Lapham, who issued the first storm warnings on November 8, 1870, was charged with the responsibility for the Great Lakes region, but he declined a permanent appointment because of poor health. To replace him, in 1871 the corps hired civilians T. B. Maury and Cleveland Abbe; they handled all of the forecasting duties until mid-1872, when General Myer assigned five lieutenants to aid them. Abbe, special assistant to the chief signal officer, began issuing forecasts entitled "Weather Synopsis and Probabilities" based on the 7:35 A.M. observations of the weather stations in 1871. Whereas the original resolution covered only the Great Lakes and Gulf and

Atlantic Coasts, a June 10, 1872, congressional appropriations bill provided that the Signal Corps establish a "system of observations and reports in charge of the Chief Signal Officer for such stations, reports and signals as may be found necessary for the benefit of agriculture and commercial interests."[7] Thus, the weather service became national in scope.

The Signal Corps began issuing weather predictions for twenty-four hours in advance in October 1872, and in 1874 forecasts included not only the type of weather but also the wind, barometric pressure, and temperature. Local Signal Corps offices distributed forecasts to thousand of rural post offices, which posted the notices on their buildings beginning in 1873. Signal flags replaced the bulletins in 1881, and by the end of 1886 they flew daily in 290 locations. Although these early predictions were frequently incorrect, they did give the average American some help in planning daily activities that the weather might affect.[8]

John Park Finley, destined to be the corps' tornado expert, enlisted in 1877. Although the waiting list for admission to the Signal Corps contained over one thousand names, Finley had help in receiving the appointment. He presented letters of recommendation from public officials and college professors at his alma mater, the Michigan State Agricultural and Mechanical College (now Michigan State University), where he had studied meteorology and climatology in relationship to agriculture and had received a bachelor of science degree in 1873. After acceptance into the corps, Finley reported to Fort Whipple, Virginia (renamed Fort Myer in 1880) for training as assistant to the noncommissioned officer in charge of a weather station. The three-month training session consisted of courses in military tactics, signaling, telegraphy, telegraph line construction, electricity, meteorology, and meteorological observation. Finley learned to take weather readings at precise times with unfailing accuracy, to compute and encipher the weather data, and to telegraph the perfect message to the central office in Washington, D.C. After completing the schooling, he served for three months as assistant to the sergeant in charge of the Philadelphia station.[9] While he was there, William Blasius's book *Storms: Their Nature, Classification, and Laws* caught his attention and motivated him to begin a systematic study of tornadoes.[10]

In accordance with the Signal Corps' custom to send an observer to survey locales that had suffered great tornado damage, Finley visited areas throughout the Central Plains in May 1879. His report, appearing in the *Report of the Chief Signal Officer* for 1880 and later as a Signal Service professional paper, suggested that the corps establish a station manned by a special observer during May, June, and July, the peak months for severe thunderstorm and tornado activity. Finley proposed Kansas City, Missouri, as the site; seventy-five years later the Weather Bureau took his advice and located its severe weather forecast center there.

General Myer died in 1880, and his replacement, General William B. Hazen, established a research unit called the "Study Room" in 1881. Finley, promoted to private first class in 1879 for his impressive reports, received permission to continue the work begun in Philadelphia on his collection of tornado reports spanning the years 1794 through 1881. The resulting report, *Character of Six Hundred Tornadoes*, was the most comprehensive study of the climatology of tornadoes to that date and became the basis for his belief that he could develop a viable system of forecasting these storms based upon rules deduced from his data. When the corps assigned him the task of testing the feasibility of forecasting tornadoes, Finley moved his "tornado studies" headquarters to Kansas City in April 1882, and during that spring he traveled extensively in Arkansas, Missouri, Kansas, Nebraska, Iowa, Illinois, and Michigan enlisting tornado spotters or "tornado reporters" for his network. The spotters, often two or three in the same county, were concentrated in the states with the highest tornado frequency. In return for their volunteer efforts to collect and report storm data, which included instrumental observations, photographs, diagrams, illustrations, and charts, the spotters received the corps' tornado publications and necessary instructions for recording and mailing their reports to the chief signal officer. Throughout 1883 and early 1884 Finley, though again stationed in Washington, D.C., continued to increase his volunteer network until the number reached 957 by June 1884. The effectiveness of the spotters' reports is reflected in the expanded length of the "Local Storms" section of the spring 1885 issue of *Monthly Weather Review.*[11]

Finley's observations and data collection led him to believe he could forecast conditions that might lead to a tornado's appearance. He realized that the successful forecast of an approaching twister would not prevent the destruction of buildings, crops, or machinery, "but it will give opportunity for preparation to protect such destruction of life, and much personal property, the latter in the shape of valuables, that are readily moved from place to place." A corollary to the preservation of life was the alleviation of anxiety and excitement in tornado-prone areas of the country, where factories and businesses often closed at the first sign of a black cloud on the horizon.[12] The same reasons for tornado forecasting—protection of life and alleviation of fear—motivated meteorologists more than one hundred years later.

Finley did not arbitrarily decide a tornado would appear on a particular day. He had developed stringent rules based on his personal observations in tornado country, statistical data gathered from the tornado observers, and historical records. In an 1884 article in *Science,* he reported the principal results of his studies, which in actuality were the first list of rules for tornado forecasting. According to Finley:

1. There is a definite portion of an area of low pressure within which the conditions for the development of tornadoes is most favorable, and this has been called the dangerous octant.

2. There is a definite relation between the position of tornado regions and the region of high contrasts in temperature, the former lying to the south and east.
3. There is a similar definite relation of position of tornado regions and the region of high contrasts in dew-point, the former being, as before, to the south and east.
4. The position of tornado regions is to the south and east of the region of high contrasts of cool northerly and warm southerly winds—a rule that seems to follow from the preceding and is of use when observations of temperature and dew-point are not accessible.
5. The relation of tornado regions to the movement of upper and lower clouds has been studied and good results are still hoped for.
6. The study of the relation of tornado regions to the form of barometric depressions seems to show that tornadoes are more frequent when the major axis of the barometric troughs trends north and south, or northeast and southwest, than when it trends east and west.[13]

In 1886 Finley restated these rules in the *Journal of the Franklin Institute* and added key surface weather map features that the forecaster must study to adequately predict the occurrence of a tornado:

1. Barometric Trough. Region. Ratio of Axes. Pressure. Departure from Normal.
2. Central Area of Barometric Minimum. Region. Pressure. Departure from Normal.
3. High Contrast of Temperature. Region. Gradient.
4. High Contrasts of Cold Northerly and Warm Southerly Winds. Region.

5. High Contrasts of Dew-point. Region. Gradient.
6. Heaviest Lower Cloud Formation. Region. Kind.
7. Opposing Movement of Lower Clouds. Region. Directions.
8. Coincident Movement of Upper and Lower Clouds. Region. Direction.
9. Opposing Movement of Upper and Lower Clouds. Region. Direction.
10. Opposing Movement of Lower Clouds and Winds. Region.[14]

Finley based his forecasts on a combination of the surface map plots and climatological data. He began issuing regular tornado predictions on an experimental basis on March 10, 1884. During March and April he posted an eight-hour prediction based on the 7:00 A.M. weather map and a second prediction based on the 3:00 P.M. one. In May, June, and July he made only one daily prediction, of sixteen-hour duration, based on the morning map. Finley divided the country between the 77th and 102nd meridians into eighteen districts, then subdivided each district into four equal parts. A prediction could cover an entire district or any part of one and had a dual nature—it advised whether conditions were favorable or unfavorable for tornado development.[15]

Were these initial attempts to forecast tornadoes successful? Finley reported in the *American Meteorological Journal* in July 1884 that he had attained a 94.29 to 98.56 percent degree of success for the months March, April, and May (table 3). He explained that "in no instance where it was predicted that conditions were favorable for the development of tornadoes did violent storms fail to occur, either hail, hurricanes, or tornadoes." Lead time for all of the destructive tornadoes was at least five hours. No prediction was entirely successful unless the characteristic funnel-shaped cloud appeared, the tornado track was within the district for which the prediction had been made,

TABLE 3

Finley's Tornado Predictions and Verifications

Month	Number of Predictions Favorable for Tornadoes	Number Verified	Number of Predictions Unfavorable for Tornadoes	Number Verified	Total Number of Predictions	Total Number Verified	Percentage Verified
March	43	6	728	721	771	727	94.29
April	25	11	909	906	934	917	98.17
May (8 hr.)	10	8	548	542	558	550	98.56
May (16 hr.)	22	3	518	511	540	514	95.19

NOTE: Finley's table did not contain the percentage verified. An apparent error in his table has been corrected. Finley listed the total number of predictions for May 16-hour period as 549.

SOURCE: John P. Finley, "Tornado Predictions," *American Meteorological Journal* 1 (July 1884): 86.

and the storm occurred within the designated eight- or sixteen-hour period.

Several individuals questioned Finley's statistical methods. G. K. Gilbert acknowledged that Finley's work showed "encouraging progress," but he questioned Finley's method of counting nonoccurrences on equal footing with tornado occurrences. When Gilbert used only actual tornado occurrences to recalculate Finley's success quotient, the result was 23 percent. He also determined that ignoring weather maps and using only climatological data would yield a 98.18 percent verification rate.[16] Others, such as G. E. Curtis, objected to Finley's use of fixed districts. Finley had reported a 100 percent verification rate for several districts, such as district I, which encompassed upstate New York and New England. Tornadoes in this area from March to May were extremely rare, so Finley had an excellent chance of being right if he predicted no tornado occurrences. In a similar manner, Finley could hardly miss if he predicted that a tornado would occur in May somewhere within district XII, which included north Texas and the Texas Panhandle. Curtis suggested that Finley use a moveable forecast area, perhaps four hundred by six hundred miles in size, which he could employ where tornadoes were most likely to occur on any given day.[17] From 1884 to 1893 several other papers addressing Finley's tornado verification data appeared in scientific and meteorological journals.[18]

One who questioned Finley's identification of tornadoes rather than merely his statistical methods for determining his tornado forecasting success was Dr. Gustavus Hinrichs, the director of the Iowa Weather Service. In an invited article in the November 1888 *American Meteorological Journal*, Hinrichs argued that many Iowa storms that Finley had classified as tornadoes were in all probability squalls or what Hinrichs called *derechos*, "straight-blow of the prairies." The Iowa weatherman was disturbed by what he perceived as a conspiracy between sensational journalists and the Signal Service to picture Iowa as a place with excessive tornadoes in order to frighten people from settling in the state and to coax those who lived in Iowa to purchase tornado insurance. To counteract Finley's claims that during the thirteen-year lifetime of the Iowa Weather Service (1876–1888) 107 tornadoes had occurred in the state, Hinrichs compiled his own list, which showed that only seven multiple tornadoes, nine single tornadoes, and fourteen minor or doubtful tornadoes had occurred. Hinrichs concluded his article with a statement that it was a "matter of duty to science and to our people to wipe out that immense body of fiction and error which forms the bulk of the tornado lists issued by and under the authority of the Signal Service" (an obvious reference to Finley's work) and requested those who dominated the field of meteorology and received government aid to "stop the manufacture of dire tornadoes, and thus free their published records from the stain of absolute untrustworthiness that now makes them useless."[19]

Finley supporters were not quiet on the subject of his perceived success at forecasting tornadoes. The October 31, 1885 *Scientific American* contained a letter from William A. Eddy, a Signal Corps tornado reporter from New York, who reported that 3,201 of the 3,228 forecasts for no tornado occurrences Finley made in 1884 were correct. Predictions for tornado occurrences were not as successful, though. From April to June 1884 only eighteen of the thirty-eight forecasts were verified, but in 1885 Finley's verification rate increased when tornadoes appeared fifteen of the nineteen times he had predicted them. In every instance when Finley had forecast tornadoes, a violent storm occurred. Eddy blamed the failure of some predictions on inadequate reports from sparsely settled regions. His report concluded with an appeal to Congress to finance a system of tornado flags or disks at every telegraph station throughout the Great Plains, the Midwest, Georgia, and South Carolina during the spring and summer months.[20]

Because the work was experimental, these initial forecasts did not reach the public, but by 1885 Finley's perceived success and his belief that these predictions should appear in the official releases of the Signal Corps created a change in official policy. The chief signal officer agreed to include a special warning when violent storms were possible, but he forbade the use of the word *tornado.* Official instructions for 1886 included the provision for notification of the directors of weather services in Minnesota, Ohio, and Alabama if tornadoes threatened their states.[21]

Almost as abruptly as they had begun, tornado forecasts ceased. Finley's tornado work was a casualty of the conflict over control of the weather service. Civilians and other military units had criticized the Signal Corps' weather unit from its inception. The Corps' disbursement officer, Captain Henry Howgate, went to prison in 1881 for embezzling ninety thousand dollars. Businesses that relied on accurate weather information, such as the Chicago Board of Trade, petitioned Congress for a better organization with more highly trained meteorologists under civilian control. Resistance within the corps' ranks to General Hazen's disciplinary methods and congressional questioning of the expenditures and lack of scientific training led in 1884 to the creation of the Allison Commission to investigate the Signal Corps. The commission's final report, issued in 1886, urged the placement of the weather service under the jurisdiction of the War Department and the closure of the training center and Study Room at Fort Myer. The army, dissatisfied with the corps' insistence that the corps should enjoy the status of a separate unit with freedom to control its own actions, feared that the corps' military service would become secondary to its weather duties in time of war.[22]

Finley, who had testified before the Allison Commission that he spent seventeen hours a day on military duties, classroom training, and tornado studies, lost favor with Hazen. The general ordered Finley's Tornado Studies Project moved to Washington, D.C., in 1885, and after a few months Hazen reassigned him to the Meteorological Records Division, where his job was to check the accuracy of weather data. Finley took command of the Signal Corps station in New York City in November 1886, but he returned to Washington the next month when Hazen became ill.

The new chief signal officer, General Adolphous W. Greely, had difficulty handling the continuing feud over control of the weather service and responded to the pressure with reorganization. Following the Allison Commission's recommendations, he closed the Study Room in March 1887 and relegated Finley to the Records Division, where Finley would no longer participate in weather forecasting. The Tornado Studies Project existed only nominally as a section of the Records Division. The June 1887 Records Division report contained only two items relating to tornadoes: the number of tornado reporters throughout the country (2,376) and the mention of routine compilation of tornado data.[23] The demise of Finley's tornado forecasting plan appeared officially in the *Report of the Chief Signal Officer* for 1887, which stated that "it is believed that the harm done by such a prediction would eventually be greater than that which results from the tornado itself."[24] Although the corps' official view was that tornado forecasts would lead to public panic (a belief that lasted well into the twentieth century), Finley's Tornado Studies Project was undoubtedly a casualty of the civilian-military feud over control of weather forecasting and research. Finley's final report on his project listed his accomplishments: twenty- three tornado circulars, four professional papers, over fifteen hundred tornado reporters, and twelve months of tornado predictions.

Although Finley was no longer officially a tornado forecaster, his book *Tornadoes: What They Are and How to Observe Them; with Practical Suggestions for the Protection of Life and Property* appeared in 1887. This composite of his previous publications included pictures, charts, maps, a list of 143 tornado characteristics, instructions for observing storms, and twenty-three tables. Among the tables were the total number of tornadoes observed from 1682 to 1886, the month of occurrence and relative frequency by state, hour of occurrence, temperature before and after the tornado, direction of storm movement, and form of the tornado cloud. Of a more practical nature were the sections on "premonitory signs," local observations that could indicate a tornado's development, and advice on seeking protection from tornadoes.

Finley urged citizens in tornado-prone areas to become their own tornado forecasters by paying particular attention to the development of peculiar clouds on the western horizon on sultry, oppressive days from April 1 through the end of September. These ominous clouds that appeared in the southwest and then almost immediately in the northwest or northeast were "unlike any ordinary and usual formation. The dark clouds at times present a deep, greenish hue which forebodes the greatest evil and leaves one to imagine quite freely of dire possibilities." The gathering clouds might appear jet black or roll "like the smoke from an engine or locomotive burning soft coal." Another significant feature of the storm clouds that could indicate tornado formation was their "pell-mell" movement from opposite directions. The funnel-shaped tornado cloud, the roar of which would signal its approach, would appear from the southwest section of the cloud mass. Finley noted that the noise of the tornado was so distinct that it alone should provide the warning citizens needed to take cover.[25]

To provide protection for humans and valuables, Finley urged citizens of tornado country to have a tornado cave or dugout located where they could get into it quickly. Those who lacked such retreats should go into the cellar and get as close to the west wall as possible.[26] He also included instruction on how to flee a storm: "*never run toward the storm nor with it*, either run to the northward or southward at a right angle from it, giving the benefit of doubt in favor of the storm."[27]

One Finley supporter was Edward S. Holden. After reading Finley's *Character of Six Hundred Tornadoes*, Holden devised a system for warning towns of approaching tornadoes. He believed that the Signal Corps had the capability to send out tornado "warnings" (actually forecasts) for a large region, perhaps an entire state, a day in advance, but because the area in question was so large, some type of local warning would be needed. Holden proposed to warn each household in a small town by a continuous ringing of a bell when a wind of destructive force (he proposed seventy miles per hour) was approaching. The five-minute lead time would give those in the storm's path adequate time to seek shelter and probably save their lives. To counteract the objection that a person might take shelter only to find that the storm was not a tornado but a strong gale, Holden wrote that because violent storms would not occur more than once a year "it would seem that one could afford to be frightened as frequently as this for the sake of immunity from an occasional tornado."

Holden envisioned an arc of telegraph poles at a distance of two to two and one-half miles to the town's south, southwest, and west. A single wire connecting the poles would end at the telegraph office, and additional telegraph wires would connect houses within the town to the office. A battery at the office would send a constant current over the wire. A magnet at each house would hold a detent that would prevent a bell from ringing as long as the circuit remained unbroken. In the event of a strong wind or tornado, the line connecting the poles outside of town would break and the bell would ring in each house. In addition, a "simple device" could fire a cannon to warn people in the fields and streets to take shelter. In a large town the circuit could end at the local fire station, which could then forward the warnings. Holden emphasized that in light of the great loss of life to tornadoes in 1883 (270 deaths occurred that year before the publication date of his article) communities should consider this cheap system of local warning.[28]

The ultimate insult to Finley came in 1890 when the *Report of the Chief Signal Officer* contained the statement, "Impressed with the number and violence of destructive tornadoes during the past year, it is believed that an investigation of phenomena of this kind on their numbers, area devastated, lives lost, and other such information might be of current interest. This work was intrusted to Professor H. A. Hazen, who has given much time and attention to these phenomena." The *Report* mentioned neither Finley's previous study on tornado prediction nor his statistical work.[29] Henry A. Hazen had graduated from Dartmouth in 1871 and had studied one year at the Thayer School of Civil Engineering before serving at the Sheffield Scientific School as instructor in drawing and assistant in physics and meteorology to Elias Loomis. Cleveland Abbe recommended his appointment as a statistician in the corps' Study Room in 1881. Hazen, a civilian employee of the corps and a second cousin of General Hazen, had served as weather forecaster and editor of *Monthly Weather Review* for a time before taking over the tornado project.

Hazen's book *The Tornado* appeared in 1890. More theoretical than Finley's volume, it summarized Hazen's views on tornadoes, including their forecastability. He believed that because the storms were "exceedingly rare" and very localized, the best predictions would have to cover thousands of square miles. Citizens of the forewarned districts should not be disturbed but should give adequate attention to the appearance of threatening clouds. As to the question of whether it would be better to omit tornado prediction entirely, Hazen stated his belief that "if the right view be taken of it, that it is a warning to look out, and not a positive statement, no one should be unduly disturbed." He hoped that in the future meteorologists would have a better understanding of conditions a few thousand feet above the ground, enabling them to issue better predictions, but he believed it was impossible to issue pinpoint forecasts for destructive tornadoes such as the one that struck Louisville, Kentucky, in 1890. Rather, he said, "all that we can do is to predict a disturbed region."[30] The Signal Office had issued predictions for severe local storms throughout the lower

Ohio Valley region nearly twelve hours in advance of the Louisville twister.

Finley's belief that meteorologists could predict tornadoes and Holden's idea that an alarm system could warn communities of approaching tornadoes disappeared for sixty years. Meanwhile, the responsibility for weather forecasting in the country shifted from the military to the civilian sector. Congress, dissatisfied with the Signal Corps' internal disharmony and lack of financial accountability, agreed with President Benjamin Harrison that the national weather service should be under civilian control. The president signed the act transferring the job of the nation's meteorologist from the Army Signal Corps to the Agriculture Department on October 1, 1890.[31]

Mark W. Harrington, professor of astronomy and director of the observatory at the University of Michigan, took control of the new civilian United States Weather Bureau; however, a lengthy feud between Harrington and Secretary of Agriculture J. Sterling Morton led to the appointment of Willis L. Moore as bureau chief in 1895. Moore was a severe critic of the Signal Corps' tornado reporters and their questionable methods of collecting data. He had worked his way through the Signal Corps ranks to become a local forecaster at Milwaukee and then the official in charge at the Chicago Weather Bureau Office. The new chief did not propose adoption of any new tornado counting methods but ordered Alfred J. Henry, chief of the Division of Records and Meteorological Data, to review the tornado reports from 1889 to 1896 and to adjust the statistics to reflect only death and destruction by actual tornadoes, not all windstorms.[32] Moore believed that "in almost all cases of great disaster there is a pronounced tendency to exaggerate the actual facts" and blamed journalists and tornado insurance companies for inflating statistics. Unquestionably, the United States had suffered great loss of life from tornadoes (1,458 deaths) during the eight years under consideration, but Moore hoped to dispel the idea that the frequency or severity of tornadoes was increasing.

Henry's report in the *Report of the Chief of the Weather Bureau 1895–1896* included the date, place, time, path length and width, direction of movement, casualties, and property losses for each tornado.[33]

Three 1899 *Monthly Weather Review* articles by editor Cleveland Abbe reflected the state of tornado study during the period. In April Abbe asserted that the number of tornadoes was not increasing; instead, the proliferation of telegraph lines and daily newspapers had brought the storms to the public's attention.[34] A second article in the same volume addressed the lack of a warning system. A *Chicago Tribune* article questioned whether the Weather Bureau could have warned Newtown, Missouri, of an approaching tornado that killed twelve on April 27, 1899. The *Tribune* realized that the bureau could not have warned Kirksville, Missouri, because the tornado formed immediately outside town, but the newspaper asked whether it was not possible through the use of telephones and telegraphs to warn others in its path. In reply Abbe gave four reasons for not issuing warnings: (1) lack of knowledge of which way the tornado would move might precipitate warning the wrong town; (2) tornadoes dissipate at will; (3) telephone operators, concerned for their own safety, would fail to forward the message to the next town; and (4) three-fourths of the tornadoes would slip unobserved between far-flung telegraph and telephone stations. Abbe explained that to be of any use, stations would have to be within a mile of each other to catch every tornado and determine its path. He did urge that large cities with adequate telephone lines, such as Saint Louis, Chicago, New York, Boston, Philadelphia, Detroit, Buffalo, Baltimore, Washington, and New Orleans, begin a serious effort to establish such a warning system. He concluded that the chance of being injured by a tornado was so slight, only about once in every ten thousand years, that the Weather Bureau had "no right to issue numerous erroneous alarms. The stoppage of business and the unnecessary fright would in its summation during a year be worse than the storms themselves." In addition, because "the certainty of destruction is absolute when the tornado comes, then it follows inevitably that there is no material advantage to be derived from any, even the most perfect, system of forewarnings and attempts at protection."[35] These statements from one of the country's leading meteorologists and editor of *Monthly Weather Review* should have generated a public outcry or at least a debate within the meteorological community, but the literature gives no indication of any response to Abbe's words. The large death tolls from recent tornadoes, especially in major cities such as Louisville, where seventy-six died on March 27, 1890, and Saint Louis, where 255 died on May 27, 1896, should have led the Weather Bureau to try anything to reduce the loss of life. Congress created the federal weather service to give notice of approach and force of storms, and the bureau was not fulfilling the law. The bureau was also insulting the public's intelligence by assuming it would panic and not know what actions to take. Regardless of the public's reactions to tornado forecasts or warnings, the bureau had the responsibility it has always had—to save as many lives as possible.

Abbe's third article addressed the concern of the Weather Bureau's Oklahoma section director, J. I. Widmeyer, who wrote on May 12, 1899, that through their tornado predictions long-range forecasters were causing unnecessary alarm among Oklahomans. Although no tornadoes had occurred in the state that year[36] and the Weather Bureau had not issued a single severe local storm forecast, Oklahoma residents were fleeing to caves and cellars whenever rain clouds or thunderstorms appeared. Widmeyer believed that exposure to the shelters' dampness resulted in more deaths than the tornadoes had ever caused and that the constant fear of the storms induced nervous troubles. Abbe agreed that the Weather Bureau should do everything possible to allay the population's fears. He blamed sensational writers of earlier decades who had painted distorted pictures of the country's severe weather for alarming the public and for labeling Kansas, Iowa, and surrounding areas "tornado states." Newspapers and residents of the plains acknowledged that they had occasional twisters, cyclones, whirlers, hailstorms, or hurricanes, but never tornadoes. Abbe concluded with a statement that tornadoes really caused no more destruction than lightning, high winds, hailstorms, droughts, or floods did, so citizens should not stop their daily routine and flee for shelter until they saw the funnel cloud approaching.[37]

Former Weather Bureau chief Harrington calculated (inaccurately) the odds of seeing a tornado or being injured by one. He wrote in *About the Weather* (1909) that tornadoes were quite uncommon, numbering only about fifty per year. By his count, the prospect of an individual in the tornado area (the Mississippi Valley east of the Great Plains) seeing a tornado in any one year was 1 in 625,000, and the chance of being injured by a tornado during a one-hundred-year lifetime was only one in a million, odds definitely not worth worrying about. Those who lived outside the tornado area had no reason to be concerned about tornadoes; they would never see one. Because of the minute chance of injury from a tornado, he reasoned, no need for warnings existed.[38]

Regardless of a person's chances of seeing a tornado or being injured by one, the responsibility of the Weather Bureau was to warn of approaching storms. Abbe failed to note that the bureau *was forecasting* the severe thunderstorms that brought lightning, high winds, hail, and torrential rains, and considering that 1,822 Americans died in tornadoes during the 1890s, the idea that people should not be concerned about a tornado until they saw it approaching was irresponsible advice. The very least the bureau should have done was to inform the population of what safety measures to take in the event of a tornado.

During Moore's tenure (1895–1913) the meteorological establishment exhibited little interest in tornadoes. None of the 131 papers presented at national meetings of Weather Bureau officials in 1898, 1901, and 1904 addressed severe storms or tornadoes. Accounts of intense storms might occasionally appear in the *Monthly Weather Review* or the *Monthly Weather Summary* of states, but meteorologists appeared to have reached a general consensus that forecasting tornadoes would do more harm than good.[39] Continuing the precedent set by the Signal Corps, the Weather Bureau Stations Regulations of 1905 contained the statement, "Forecasts of tornadoes are prohibited." The ban was reiterated in the 1915 and 1934 regulations. When conditions were favorable for tornado formation, district forecasters could use the term *severe thunderstorm* or *severe local storms,* but only the chief of the Weather Bureau or, in his absence, the chief of the Forecast Service could use the phrase "conditions are favorable for destructive local storms."[40] The restriction remained in place until 1938. Although the Weather Bureau may have been justified in not having a national tornado forecasting system in the nineteenth century because meteorologists lacked crucial data such as upper air soundings, it could have allowed local forecasters, who were familiar with their region's predilection for tornadoes, to use the word *tornado,* especially during the 1920s and 1930s, the two deadliest decades in the nation's history.[41] One cannot help but ask what harm forecasts would have done, but maybe the more important question is, How many lives would forecasts have saved?

During the period when the Weather Bureau did not issue tornado forecasts, citizens of tornado-prone areas learned to rely on their senses, observations of nature, and *The Farmer's Almanac* for weather predictions. Folk wisdom had taught plains residents that the sky would often turn green, the wind would cease blowing, and animals would become agitated just before a tornado. The appearance of any of these signs or the roar of the wind would alert the family's weather observer, usually the father, to lead the family into the underground storm cellar located near the house.[42] One Apache method of weather forecasting was reading patterns in bear grease. According to this theory, animal cells respond to the weather even after the animal has died. The bear grease in animal bladders or jars formed various patterns on the sides of the containers, and reportedly tornadoes would follow observations of funnel shapes that appeared in the grease.[43]

Research on tornadoes came to a virtual halt during the time of the ban on the word *tornado*. Europeans were especially surprised that in the United States, the home of the majority of the world's deadly tornadoes, the literature consisted only of compilations of statistics and descriptions of damage.[44] European meteorologists produced the few articles on tornado formation and prediction that appeared in American meteorological journals during the first two decades of the twentieth century. Frenchman E. Durand-Gréville wrote in 1914 that "tornadoes always originate on the front edge of the squall-zone." Because these squall zones (or squall lines) in the United States advance from west to east, the central bureau in Washington, D.C., could plot storm reports received by telegraph on maps and could predict the squall's passage over a particular location several hours in advance. When the front edge of the squall had passed, the threat of tornadoes would be over. Durand-Gréville admitted that France had not yet adopted his method but said he would be "very happy if the application of the law of squalls and tornadoes were to be made in a country which tornadoes seem to have selected as the land of their predilection." Although forewarning could not divert the storm's fury, it could save many lives.[45]

Had the American weather establishment heeded Durand-Gréville's suggestion and instituted some type of warning system, perhaps the effects of the greatest tornado disaster in the country's history, on March 18, 1925, would have been mitigated. A warm, moist layer of Gulf air covered the Midwest corn belt that day, but a cold air mass from Canada was rushing southward. Citizens of Illinois and Indiana were looking forward to the approaching spring. No one, not even the Weather Bureau, realized the imminent danger. The bureau predicted

thunderstorms in Indiana, Illinois, and Kentucky for that afternoon. There was no mention of tornado danger because regulations prohibited such a prediction, but neither was there the dreaded allowable forecast, "conditions are favorable for destructive local storms." At 1:01 P.M. the snapping of trees at Ellington, Missouri, heralded the beginning of three and one-half hours of the most intense tornado destruction in the country's history. By day's end 695 citizens of Missouri, Illinois, and Indiana were dead, including 234 in Murphysboro, Illinois. Injuries numbered over two thousand. Estimated property losses totaled $16.5 million. The tornado established all-time records for path length (219 miles) and ground speed (an average sixty-two miles per hour). No warning system could have saved the property, but many lives could have been saved. Some tornadoes do not follow a predictable path, but this was not true of the Tri-State Tornado. For 183 miles the tornado followed a slight topographic ridge and maintained an exact heading. The appearance of the storm—a boiling mass of clouds about three-quarters of a mile wide rather than the typical visual funnel—did not give people enough time to react.[46] However, the question remains: Could a warning to the communities in the tornado's path have saved lives?

Why did the Weather Bureau pay so little attention to tornadoes before World War II? One explanation may be that the bureau, under the jurisdiction of the Department of Agriculture, was most interested in providing farmers with weather information that might affect crops, and tornadoes caused little crop damage. Another reason may be the general attitude of the government that tornado-prone areas did not merit much attention because they were rural and sparsely populated. The bureau's belief that tornado forecasts might cause fear or panic may reflect the feeling that farmers and inhabitants of small towns in the Plains and the South were uneducated and unable to understand the difference between a forecast and a warning. The most likely explanations, though, are scientific and technological.

Although weather observations and data collecting had been around for centuries, during this period meteorology was a relatively new science, and most Weather Bureau personnel had learned their craft through on-the-job experience rather than in a college classroom. Only in the mid-1930s did the bureau adopt the frontal theory of storms and hire university men trained in the use of this new forecasting method. In addition, meteorologists lacked some necessary components for adequate tornado forecasting, especially upper-air soundings, to provide a profile of the atmosphere up to forty-five thousand feet. More noticeably, the country lacked an adequate communications system to spread notice of impending danger. Regardless of the reasons, failing to warn those who could be notified in some manner had devastating consequences. Even if the forecasts and warnings were imperfect, some of the 4,151 Americans who died in tornadoes from 1920 to 1939 might have survived.

In spite of advancements in meteorology and technology, a system of tornado forecasting and warnings was as nonexistent in 1940 as it had been in 1870. Two events of the next decade, the transfer of the Weather Bureau to the Department of Commerce and World War II, would bring substantial changes to the civilian weather service and lead to a program to warn the public of impending danger from nature's most violent windstorms.

CHAPTER 3

Wartime Warnings and Military Forecasting

When Congress created the Weather Bureau to aid the nation's farmers in 1890, it could not foresee that aviation would depend on accurate weather forecasting even more than agriculture did. The bureau began issuing special bulletins and forecasts for domestic military training flights and the Post Office's airmail carriers in 1918. By 1920 the bureau had established flight forecast centers for the U.S. Army and Post Office in Washington, D.C., Chicago, and San Francisco, but the Air Commerce Act of 1926 obligated the bureau to expand its services to the commercial segment as well. This statute required the chief of the Weather Bureau to provide "weather reports, forecasts, warnings, and advices as may be required to promote the safety and efficiency of air navigation in the United States . . . and observe, measure, and investigate atmospheric phenomena." To fulfill its mission, the bureau established weather observation stations along flight paths to provide weather data and airport stations to give pilots preflight briefings. Airport stations increased from fifty in 1930 to one hundred in 1941. Many observers were either military personnel or employees of the Bureau of Air Commerce (predecessor of the Federal Aviation Administration). By the early 1930s, airport control towers were broadcasting forecasts every thirty minutes.[1]

The 1938 law that created the Civil Aeronautics Authority reiterated the 1926 law's requirements, but the Weather Bureau was still under the jurisdiction of the Department of Agriculture, which did not have a vested interest in fulfilling Congress's instructions. As commercial aviation expanded dramatically in the 1930s, so did the need for more timely and accurate weather forecasts. To alleviate the situation, Reorganization Plan IV of June 30, 1940, transferred the Weather Bureau to the Department of Commerce; almost immediately, however, the needs of war production supplanted those of commercial aviation and the general public, and the Weather Bureau became part of the country's war effort.[2]

Weather can often determine the victor in a military battle or campaign. The Spanish Armada in 1588 and Napoleon in 1812 learned this lesson the hard way, but meteorologists did not become significant members of the military establishment until World War I, when the inability of the Weather Bureau to meet the military's demands led to the creation of military weather services. When war began in Europe in 1939, twenty-two officers and 180 enlisted personnel staffed the Army Air Corps Weather Services' forty weather stations.[3] Representatives of the U.S. Army, Navy, and Weather Bureau formed the Defense Meteorological Committee in 1941 to facilitate communications between the military and civilian weather establishments, and President Franklin Roosevelt issued Executive Order 8991 on December 26 of the same year, designating the Weather Bureau a war agency.[4]

When the United States entered World War II on December 8, 1941, the Defense Meteorological Committee became the Joint Meteorological Committee of the U.S. Joint Chiefs of Staff. One of the committee's first actions was to initiate plans to prevent weather information from reaching enemy hands. Because weather fronts move eastward across the United States and the North Atlantic toward Europe, the United States had the advantage of knowing weather conditions in the Atlantic throughout the war. In spite of persistent spy efforts and utilization of submarine-placed automatic weather stations, Adolph Hitler had difficulty learning about weather conditions that might affect his navy's Atlantic activities. To prevent aiding the Axis powers, the Weather Bureau began coding all weather information broadcast on U.S. Navy and Civil Aeronautics Administration radio stations.

In early 1942 the Office of Censorship issued specific censorship codes for press and radio use, asking the media to withhold the following information:

> Weather forecasts, other than officially issued by the Weather Bureau, the routine forecasts printed by a single newspaper to cover only the state in which it is published and not more than four adjoining states, portions of which lie within a radius of 150 miles from the point of publication.
>
> Consolidated temperature tables covering more than 20 stations, in any one newspaper.
>
> Weather "round-up" stories covering actual conditions throughout more than one state, except when given out by the Weather Bureau.
>
> NOTE: Special forecasts issued by the Weather Bureau warning of unusual conditions or special reports issued by the Weather Bureau concerning temperature tables, or news stories warning the public of dangerous streets or roads within 150 miles of the point of publication, are all acceptable for publication.[5]

The Weather Bureau took censorship of weather reports very seriously. Instead of detailed forecasts, the bureau issued vague statements such as "tomorrow will be warmer" or "today will be a good day to cut hay" and published weather maps only after they were more than a week old. Censorship even applied to private citizens. When First Lady Eleanor Roosevelt mentioned rain in California in her newspaper column, the Office of Censorship reprimanded her.[6]

According to Circular Letter 77-40 of December 6, 1940, one of the aims of the Weather Bureau was "to issue and promptly distribute to the public adequate and timely warnings covering unusual and dangerous weather conditions." Senior forecasters at bureau offices were responsible for issuing warnings of severe weather conditions, such as a "severe local wind storm warnings (equivalent to tornado warnings)" in their districts.[7] This service was extremely valuable to the munitions industry. As early as 1940, American factories were manufacturing war materiel, especially gunpowder and bombs. Many of the ordnance plants that produced these highly explosive materials were located in the central and southern states, areas highly prone to severe thunderstorms and tornadoes. At plant officials' request the Weather Bureau organized a Severe Storm Warning Service to issue warnings of approaching thunderstorms. In response, plant supervisors would determine whether or not to evacuate their workers. The greatest fear from these storms was lightning-generated fires and explosions. The Warning Service would issue an "all clear" once the storm had passed, enabling employees to return to work immediately, thus limiting the loss of production. The knowledge that supervisory personnel and the Weather Bureau were looking out for their safety raised employee morale.[8]

Volunteer observers stationed at several points within a thirty-five-mile radius of an ordnance plant staffed the storm warning system. Observers telephoned reports of approaching thunderstorms to a central Weather Bureau station or a plant office, where a meteorologist or designated official interpreted the reports and made recommendations for plant evacuation. For munitions plants located in sparsely populated areas, a volunteer spotter network was impractical. In such cases, supervisory personnel patrolled the area in radio-equipped trucks and transmitted information on thunderstorms or other hazardous weather conditions to the plant. By December 1942 a severe storm spotter service was in existence around more than one hundred munitions manufacturing plants and storage facilities.[9]

The warning's success varied from location to location. The Iowa Industrial and Defense Commission reported that the spotter network had reported the seventeen storms that occurred during the summer of 1944 in ample time to warn plants. The Iowa commission went beyond the service's original intention by putting the storm warnings on the teletype and broadcasting them to planes in flight as well as to regional flight control centers. The storm spotting network around the Bluebonnet Ordnance Plant at McGregor, Texas, alerted the plant 150 times in 1943; six times the storm's severity required plant evacuation. In contrast, the Report Collection Center at Amarillo, Texas, reported that the network proved of little value in the Texas Panhandle because of limited telephone service in the area.[10]

Munitions plants were not the only entities that needed severe weather warnings; aircraft were especially susceptible to windstorms. The U.S. Navy Department of Aeronautics issued a directive on tornadoes in March 1943. The report acknowledged that tornadoes were impossible to forecast because of their highly localized nature but that if naval station personnel had adequate warning of an approaching tornado, they could help pilots fly their aircraft to safety. If time were limited, crews could secure the planes in hangars and remove from the area all objects such as rocks, lumber, and gasoline and oil drums that the tornado might hurl against planes and buildings.[11] In light of this directive, the navy's aerological officer at Hensley Field in Dallas requested a network around the naval installations in the Dallas–Fort Worth area; the Weather Bureau complied in August 1943. The following spring, Civil Defense officials met with their counterparts from the U.S. Weather Bureau, Army, and Navy to discuss Civil Defense's participation in the Severe Storm Warning Service. At the meeting Weather Bureau personnel agreed to train and coordinate the observers Civil Defense would provide, and the U.S. Army Air Corps and Navy advised the Weather Bureau where they needed networks. An organization of seventy-five spotter networks in Oklahoma, Texas, New Mexico, Louisiana, and Arkansas began in May 1944. During the next two months the Army Air Corps requested additional observers for its military bases and airfields in the South, Midwest, and Northern Plains. By February 1945 some 162 Severe Storm Warning Service networks with 3,685 observers were in operation.[12]

Outside the military realm, the Weather Bureau instituted an experimental tornado warning program in the Saint Louis, Kansas City, and Wichita areas in the spring of 1943. Newspaper and radio announcements encouraged citizens in those locales to report tornado sightings to the local Weather Bureau Office, which would relay the warnings to local radio stations for broadcast to the public. During the first year of operation only the Wichita area observed tornadoes. The Regional Weather Bureau Office in Kansas City, pleased with the accurate reports from Wichita's public-minded citizens and the cooperation of local radio stations in warning people in the tornadoes' paths, requested that the system continue in 1944.[13] Francis W. Reichelderfer, the chief of the Weather Bureau, also praised the system.[14] His February 1945 report on the organization and operation of the severe storm warning service said, "The position and direction of movement of tornadoes on eight occasions as reported to and broadcast by the Wichita Weather Bureau has shown that public reports on such storms can be reliable and that people can react in a rational manner when radio warnings of such storms are received."[15]

Not everyone in Wichita appreciated the tornado warnings. H. M. Van Auken, general manager of the Wichita Chamber of Commerce, chastised the Wichita Weather Bureau Office for using the word *tornado* in a June 21, 1948, warning. Although the chamber members were unhappy, the warnings undoubtedly saved lives when the tornado cut an eight-hundred-yard path through a residential area, injuring twelve. Van Auken wrote, "A statement to the effect that a tornado is approaching makes news which is carried by the press to all parts of the country, creating unfavorable publicity for the city and community. We feel that the interests of all concerned may be served best by refraining from the word tornado."[16] Van Auken explained that tornadoes were rare in Kansas, and the publicity from the Bureau's warnings was jeopardizing the city's industrial development. Victor Phillips, meteorologist at the Wichita Weather Bureau Office, wrote his regional director in Kansas City that the Wichita Office had never used the word "tornado" when it issued warnings for severe weather in the area, although the use of the term would not have violated Weather Bureau policy.[17] Evidently, the Chamber of Commerce was more interested in bringing money into the community than in saving lives. Van Auken's statement that tornadoes were rare in Kansas was false. From 1900 through 1947 Kansas had more than four hundred tornadoes, which resulted in 251 deaths. Sedgwick County (Wichita) suffered four tornadoes during this time, including a May 25, 1917, twister that killed twenty-three and injured seventy. On March 18, 1948, only three months before Van Auken wrote his complaint letter, a tornado touched down eight miles southeast of Wichita, injuring two and causing over two hundred thousand dollars in damage.[18]

Most of the warning networks disbanded after the war, but a few in Kansas and the upper Midwest remained in operation. Reichelderfer recognized the importance of even a few minutes' warning to those "downwind" of a tornado. His memo to the bureau's Storm Research and Forecasting staff in May 1949 called for a leader to create and supervise a network of cooperative observers who would immediately contact their local Weather Bureau Office when they spotted a severe storm that had not been forecast.[19] Kansas City and Wichita served as models for official networks, but most storm spotters in the 1940s were members of unofficial networks of farmers and townspeople who spread the alarm by siren, bell, or a special signal on rural telephone lines. The Weather Bureau gave all spotters appointment certificates and the Severe Storm Warning Service booklet containing instructions, but there were no formal training programs.[20]

The Weather Bureau relied heavily on a tornado warning program rather than tornado forecasting during the 1940s for two reasons: fear of public reaction and lack of confidence in tornado forecasting. When news of the implementation of severe storm and tornado spotter networks reached the press, rumors that the Weather Bureau would soon start issuing tornado forecasts spread like wildfire. Fearing that public confidence in the Weather Bureau would be impaired if tornadoes did not occur when they were forecast, Reichelderfer instructed his subordinates to squelch all false rumors of impending tornadoes.[21] In a June 1, 1943, letter to all Weather Bureau offices, Reichelderfer recognized that meteorologists, through the use of modern weather map analysis techniques, could recognize the atmospheric conditions conducive to tornado formation but said forecasting a tornado for a specific time and place was impractical.[22] The chief maintained this belief throughout the decade, telling the Fort Worth regional director in May 1949 that the "best scientific opinion not only among government meteorologists but also throughout the profession holds that it is impossible at present to forecast the localities and time of occurrence of tornadoes even a few hours in advance and that techniques for 'pinpoint' forecasting of these storms are unlikely to be developed for a long time, if ever." The best meteorologists could hope for was to predict twelve to twenty-four hours ahead of time a three- or four-state area in which conditions might be favorable for tornado formation. Although the forecasts would be correct about 75 percent of the time, tornadoes would probably occur in only three or four counties of the hundreds included in the forecast area. The bureau feared that general tornado forecasts might influence people to stay at home, thus adversely affecting business and industry.[23] After taking all of these factors into account, Reichelderfer decided that the Weather Bureau should not undertake a centralized tornado forecasting program but that it would allow individual Weather Bureau offices to issue forecasts only if they were certain the situation warranted such action.

At the time Reichelderfer was questioning whether tornado forecasting would ever be feasible, the U.S. Air Force had already begun a tornado forecasting program, but the circumstances behind its initiation were quite unusual. A destructive tornado raked Tinker Air Force Base in Oklahoma City at 10:22 P.M. on March 20, 1948, injuring eight people and causing $10 million in damage to aircraft. Neither the military's Air Weather Service (AWS) nor the Oklahoma City Weather Bureau Office had issued a forecast for the storm.[24]

On weather duty at the base that evening was Captain Robert C. Miller of California, who had arrived at the Tinker Air Force Base Weather Station only a few weeks earlier to serve under Major Ernest J. Fawbush. Miller had enrolled at Los Angeles Junior College in 1939 to pursue a mathematics teaching credential and had transferred to Occidental College in Los Angeles in the fall of 1941 with the intention of obtaining a degree in physics and mathematics. The entrance of the United States into World War II led Miller to enlist in the U.S. Army Air Corps (AAC) in 1942. While serving as a cook at Chanute Air Force Base in Illinois, he read an announcement about the new weather course at the AAC Technical Training Center in Grand Rapids, Michigan. The ACC would train enlisted personnel who had completed their junior year in college in the sciences or engineering and would commission them as second lieutenant ACC weather officers upon the course's completion.

Miller entered the weather course's first class in January 1943. The school's director, Colonel Don McNeal, had an M.S. in meteorology from the California Institute of Technology and stressed practical methods in meteorology. McNeal often disagreed with the University Meteorological Committee—representatives from the University of Chicago, New York University, MIT, Cal Tech, and UCLA—who oversaw the Weather Course's curriculum and emphasized theory. The cadets at Grand Rapids attended lectures by the Weather Bureau's Harry Wexler, Athelstan Spilhaus from New York University, and John Leighly of the University of California Berkeley's Geography Department for seven hours a day, six days a week. In addition, they took military and physical training one and one-half hours daily and had compulsory, supervised study two hours a day.

Upon completion of the training, Miller served two years as a weather officer in Dutch New Guinea. When the war ended, he decided to remain in the military rather than complete his degree. While serving at Fort Benning, Georgia, he developed the analytical technique that he would routinely employ at Tinker Field. Miller charted meteorological data from the surface and the 500, 700, and 850 millibar levels of the atmosphere on U.S. maps drawn on sheets of clear acetate. Each atmospheric height had its own color of acetate. When he stacked the maps on top of each other, he had a three-dimensional view of the atmosphere.[25]

Miller's California residency[26] and wartime service in the South Pacific had not prepared him for duty in a tornado-prone area. The 9 P.M. forecast for March 20, 1948, warned of possible gusty surface winds up to thirty-five miles per hour without thunderstorms, but by 9:30 thunderstorms were in progress only twenty miles southwest of the base. Miller, observing the vicious thunderstorm cells on the APQ-13 radar screen,[27] realized the storm was approaching very rapidly. The sergeant on duty typed up a warning for thunderstorms with high winds, but it was too late to secure the aircraft. At 10:20 P.M. the Weather Bureau Office at Will Rogers Airport, just seven miles southwest of Tinker Field, reported a heavy thunderstorm with ninety-two-mile-per-hour wind gusts, and, worst of all, a tornado on the ground moving northeast. Within minutes the twister passed through the air base, blowing out the control tower windows and damaging beyond repair seventeen C-54s, fifteen P-47s, and two B-29s before dissipating over the northeastern edge of the base.

The following morning five general officers arrived from Washington, D.C., to investigate the incident. Major Fawbush explained the difficulty involved in forecasting tornadoes and the reluctance of both the military and civilian weather services to issue warnings to the public. The investigators concluded that "due to the nature of the storm it was not forecastable given the present state of the art" and recommended that the meteorological community find a method of alerting the public and military establishments to approaching tornadoes in order to minimize loss of life and property.[28]

That afternoon Fred S. Borum, commanding general of the Oklahoma City Air Materiel Area, requested the AWS to have the Tinker Base Weather Station study the possibility of forecasting tornadoes. Fawbush, who had maintained an interest in tornadoes and severe thunderstorms for some years, and Miller led the investigation. For three days the men analyzed the surface and upper-air weather charts prior to the March 20 tornado, compared them with charts of conditions preceding several other tornadoes, and noted the similarities in temperature, barometric pressure, humidity, wind direction and velocity, and lifting mechanisms such as fronts that preceded the storms. Incorporating the work of Weather Bureau personnel such as J. R. Lloyd, A. K. Showalter, and J. R. Fulks with their own findings, Fawbush and Miller listed several weather parameters they considered sufficient to result in a tornado outbreak when all were present simultaneously.[29] They proposed that tornadoes develop only when the following six conditions are present:

1. A layer of moist air near the earth's surface must be surmounted by a deep layer of dry air.

2. The horizontal moisture distribution within the moist layer must exhibit a distinct maximum along a relatively narrow band (i.e., a moisture wedge or ridge).
3. The horizontal distribution of winds aloft must exhibit a maximum of speed along a relatively narrow band at some level between 10,000 and 20,000 feet, with the maximum speed exceeding 35 knots.
4. The vertical projection of the axis of wind maximum must intersect the axis of the moisture ridge.
5. The temperature distribution of the air column as a whole must be such as to indicate conditional instability.
6. The moist layer must be subjected to appreciable lifting.[30]

If all conditions were present simultaneously, a tornado would form (1) within the area where the pressure at the level of free convection was 650 millibars or higher, (2) on the windward border of the moisture wedge near the fifty-five-degree Fahrenheit dew-point isotherm, (3) where the upper dry tongue crosses over the lower moist ridge, (4) near the intersection of this axis with the windward edge of the moist tongue at low levels. Fawbush and Miller acknowledged that other meteorologists had expressed or implied some of the rules individually, but they believed application of the complete set was necessary for a successful forecast.[31] The forecaster would have to study surface and upper air data, determine the existence of all of the conditions, and project them in space and time in order to issue a "tornado threat area" announcement with reasonable lead time. The size of the threatened area would be twenty to thirty thousand square miles, and the forecasting procedure would be tedious and time-consuming.

While studying the morning weather charts on March 25, 1948, Fawbush and Miller noticed that the atmospheric conditions were virtually identical to those of March 20, the morning of the tornado. They prepared a prognostic chart for 6 P.M. After interpreting the data, they realized that central Oklahoma would be the area of primary tornado threat that evening. Immediately they notified General Borum, an active participant in tracking severe weather on radar, who shortly thereafter arrived at the base. Fawbush and Miller told him that the conditions were very similar to those that had existed before the previous tornado. The general decided that Fawbush and Miller should issue a forecast for heavy thunderstorms for Tinker Field during the late afternoon. Borum activated his new base warning system. By 2 P.M. the radar scope began picking up thunderstorm echoes along a line sixty miles northwest to one hundred miles southwest of Oklahoma City. The squall line was approaching Tinker Field at twenty-seven miles per hour, which would place it over the base near 6 P.M. The three men agreed that atmospheric conditions were so similar to those of March 20 that it seemed logical to issue a tornado forecast. Although they knew the chances of a second tornado hitting the same location within five days were less than one in twenty million, the meteorologists followed Borum's orders and issued the first tornado forecast. Within minutes Fawbush composed the message, and Miller typed it and passed it to Base Operations for dissemination. Base personnel carried out the General's tornado safety plan, which included securing aircraft in hangers, diverting incoming air traffic, and moving to places of safety.

Locations along the approaching squall line reported no hail or high winds and definitely no tornadoes. Fawbush and Miller, thinking that they had issued their forecast in error, hoped that General W. O. Senter, commander of the AWS, would treat them mercifully. At the Oklahoma City Weather Bureau Office the squall line produced only a light thunderstorm with twenty-six mile per hour winds and pea-sized hail. Believing his forecast had failed, Miller drove home. About 6 P.M. he heard distant thunder, and rain began, but there was little wind. A short time later Miller heard in passing only part of a radio bulletin about Tinker Field and a destructive tornado and wondered why the station was continuing to talk about the five-day-old event. After trying to call the base weather station and finding the phone lines dead, Miller drove to the base to determine the source of the problem. On the way he viewed with disbelief downed utility poles and debris strewn everywhere. He reached the station to find a jubilant Fawbush, who described in detail his personal observation of the tornado: As the squall line approached the airfield, two thunderstorms had merged, and the sky turned greenish black. Clouds began a slow counterclockwise rotation at the merger point, and a large cone-shaped cloud dropped to the ground. Within three or four minutes the tornado had destroyed $6 million in property but fortunately injured no one. Both General Borum's tornado disaster plan and Fawbush and Miller's forecast had been successful.

During the remainder of 1948 Miller and Fawbush devoted their time to refining and testing their basic forecast rules to determine whether they could repeat their first success. Miller searched through years of newspaper articles and reports in the Oklahoma City Weather Bureau Office and compiled data on previous Oklahoma tornadoes. William Maughan, meteorologist in charge of the Oklahoma City Office, and M. O. Asp, Oklahoma State climatologist, provided expert advice. Maughan believed in Miller and Fawbush's method but was unsuccessful in convincing his Weather Bureau superiors that tornado forecasting was feasible.[32]

The Tinker Air Weather Service Office issued its second tornado forecast for the area of eastern Oklahoma south of Tulsa and east of McAlester at 3 P.M. on March 25, 1949, the anniversary of the first successful forecast. Two tornadoes touched down just northeast of McAlester at 9 P.M. Miller and Fawbush informed Maughan of their success, and he forwarded the information to the Regional Weather Bureau Office in Kansas City. J. R. Lloyd, the meteorologist in charge at Kansas City, was, according to

Joseph Galway, "a man who set his own priorities and assumed that he needed no authorization from higher authority for whatever he wanted to do, whenever he wanted to do it."[33] Impressed with Fawbush and Miller's success, Lloyd gave Maughan permission to pass any future forecasts to the Oklahoma Highway Patrol and American Red Cross to allow them to prepare for disaster. In spite of two successful forecasts, Fawbush and Miller remained apprehensive when they issued a third forecast on April 30, 1949, for an area thirty miles on either side of a line from Altus to just south of Tulsa. Maughan notified the Highway Patrol and Red Cross, but the Weather Bureau did not forward the forecast to the civilian population. With the exception of these two agencies, AWS forecasts were provided only to military establishments. That day thirteen tornadoes touched down in Oklahoma, killing six and injuring seventy-seven. Ten of the tornadoes had occurred in or very near the forecast area.[34]

Armed with confidence, Fawbush and Miller expanded their forecasts into adjacent states. On May 6, 1949, they warned military facilities in a triangular area bounded by Amarillo, Lubbock, and Childress, Texas, that tornadoes were possible. That evening two tornadoes touched down inside the threat area. Nine days later the Tinker office issued its first two-state forecast. They notified all military bases sixty miles either side of a line from Amarillo, Texas, to Gage, Oklahoma, of possible tornadoes. That evening a destructive tornado killed seven, injured eighty-two, and destroyed 172 houses when it plowed through a World War II veteran's housing complex in Amarillo. Because they had no authority to issue their forecast to any civilian organization except the Oklahoma Highway Patrol and Red Cross, the Tinker Field meteorologists had not told the Amarillo Weather Bureau Office about possible tornadoes. When the tornado formed just west of the city, Henry C. Winburn, Amarillo Weather Bureau Office chief, took an unprecedented step: At 8:17 P.M. he broke into radio broadcasts to warn residents to take cover from a tornado approaching the city.[35]

Winburn sent a full report to Erle L. Hardy, director of the Regional Weather Bureau Office at Fort Worth, who informed Reichelderfer that problems would most likely occur over lack of a civilian forecast for the Amarillo tornado. Hardy's fears proved correct. An *Amarillo Globe* column of May 16 complimented Winburn and his staff but questioned why the citizens of the Panhandle city had no foreknowledge of possible tornado development. The columnist severely criticized the Weather Bureau's official forecast for Sunday—"Partly cloudy Sunday and Monday"—and blamed Amarillo's weather forecasting problems on bureau restrictions. The attack continued: "The weather bureau, like most government bureaus, is so scared of making a mistake that it straddles the fence. Its fear of error results in a string of continuing errors. If some observer would take the bull by the horn and forecast 'possible' tornadoes he'd be fired if the tornadoes didn't bob up. . . . If there is any one department of the government that needs a complete over-hauling more than any other it's the United States Weather Department."[36] Senator Lyndon Johnson wrote Reichelderfer to inquire why the Weather Bureau was not doing more to forecast tornadoes, and the chief responded that "the public in times of emergency such as tornado disasters, is inclined to expect more information from the Weather Bureau than it is humanly possible for meteorologists to provide." Reichelderfer explained that because tornadoes were so short lived and the exact spot where they would occur was unpredictable, the best the bureau could do was to inform the public of the approach of a general storm area in which tornadoes might occur.[37] If the columnist or Senator Johnson had known that Fawbush and Miller had forecasted the Amarillo tornado and had notified the military but not civilians that tornadoes were possible, the uproar would have been greater.

After the Amarillo tornado Reichelderfer sent a letter to all district forecast centers, criticizing them for failing to recognize potential severe weather conditions and requesting all forecasters to review the meteorological conditions under which tornadoes might occur. He also instructed them to include the possibility of severe or destructive local storms, or similar phraseology, in the forecasts when there was reasonable indication that tornadoes would develop but to be very judicious in the use of such forecasts lest the public severely criticize the bureau for issuing too many warnings.[38]

While the bureau was avoiding tornado forecasting, Fawbush and Miller continued to issue their predictions. The Oklahoma Highway Patrol leaked the Tinker Weather Office's May 20, 1949, forecast to the public. Thirteen of the fifteen tornadoes sighted in Oklahoma that day occurred within the predicted area. The tornado forecasters were confident that their methods would work well in tornado-prone areas such as Oklahoma and the Texas Panhandle, but to prove their system was universal they needed to test it outside the Plains during a time of year when tornadoes were at a minimum. The opportunity arose on November 24, 1949, when they predicted tornadoes in an area of Alabama and western Georgia. After three tornadoes occurred in the forecast area, the air force meteorologists felt confident their method would work anywhere in the country.[39]

By 1950 many civilians knew the U.S. Air Force was issuing tornado forecasts. Though official distribution of the forecasts was limited to air force and army bases and the Oklahoma City and Kansas City Weather Bureau Offices, civilian personnel on the bases would notify their families of impending severe weather, and the word spread rapidly throughout communities surrounding military establishments.

In February 1951 the air force assigned Fawbush and Miller to the newly established Air Weather Service Severe Weather Warning Center (SWWC) at Tinker Air Force Base, the first center dedicated solely to predicting severe weather and especially tornadoes. The area of forecast responsibility included the entire United States between the Appalachian and Rocky Mountains. That year the SWWC issued 156 tornado forecasts; tornadoes appeared in the forecast area 102 times.[40]

Fawbush and Miller first published their tornado forecasting method in the January 1951 *Bulletin of the American Meteorological Society*, although they had presented a paper on the subject at the AMS January 1950 meeting. By spring 1953 the tornado forecasting team had refined its forecasting techniques. In an *Air University Quarterly Review* article they acknowledged that it was impossible to forecast all tornadoes and equally impossible to foretell the precise area in which they would occur, but they could forecast the thunderstorm conditions necessary to develop tornadoes and could delineate reasonable geographical boundaries in which the phenomena were likely to occur. Because tornado forecasts required special techniques and hours of the meteorologists' undivided attention to plot numerous detailed synoptic maps, their methods were not practical for base weather stations but were limited to use in special offices such as SWWC. The SWWC forecasters reduced the number of essential elements required for tornado generation to four by eliminating condition number four and combining numbers one and five and emphasized that only when all four conditions were present simultaneously was a tornado forecast justified. To aid in the interpretation of the mass of raw data, the SWWC established a list of "Procedures for Determining Potential Instability," which became the pattern for severe weather criteria checklists at the Kansas City Weather Bureau Office.[41] Fawbush and Miller achieved great success in tornado forecasting. From March 25, 1949, to May 31, 1950, they predicted tornadoes on thirty-four occasions, and twisters formed in thirty-one of the situations with an average lead time (time between issuance of the forecast and the first tornado's appearance) of six hours.[42] For the air force, timeliness was as important as accuracy, and Fawbush and Miller had given base commanders six hours' advance notification in 27 percent of their forecasts, three to six hours' in 37 percent, and one to three hours' in 30 percent. In only 2 percent of the cases did tornadoes occur before the forecast's transmission. In the remaining 4 percent, the exact

TABLE 4

Fawbush and Miller's Tornado Forecast Verifications

	1948	1949	1950	1951	1952	1953	1954
Forecasts issued	1	14	33	87	112	233	211
Positive verification: forecast areas contained 1 or more tornadoes	1	13	29	60	65	155	133
Partial verification: times tornadoes occurred within 150 miles of forecast area	0	1	3	18	14	12	16
Tornado forecast not issued, but tornado occurred in an area where hail or wind forecasted	—	—	7	56	40	87	138
Tornado occurred, no forecast issued	—	—	—	11	12	13	13
Total number of individual tornadoes reported	1*	14*	39*	145	304	454	580
Total number of individual tornadoes occurring in severe storm forecast areas	1	14	39	134	292	411	567

*incomplete data

SOURCE: U.S. Department of Defense, U.S. Air Force, Air Weather Service, *Tornadoes and Related Severe Weather*, 4.

time of the tornado's occurrence was unknown. The AWS Headquarters published Fawbush and Miller's verification statistics for the previous seven years in 1955 (table 4) and praised the SWWC for providing excellent lead times.[43]

SWWC's tornado predictions came into question on September 1, 1952, when it forecast severe thunderstorms with wind gusts of forty to sixty miles per hour for a section of north central Texas, including Fort Worth's Carswell Air Force Base. Because it was Labor Day and so few personnel were on duty, the officers gave the forecast little attention. After all, sixty- mile-per-hour gusts were typical for the area and had caused little damage in the past. About 6 P.M. a freak tornado, producing winds estimated at 125 miles per hour, struck the base's flight line where 107 B-36s, the Strategic Air Command's massive intercontinental bombers, were parked wing to wing. The short-lived funnel damaged 106 of the 139-ton airplanes and completely demolished one. As effectively as an enemy, nature had destroyed more than half of the United States' strategic bombing capability. In the control tower the anemometer indicated ninety-one miles per hour before part of the device blew away. Radar operators had tracked the storm to the edge of the base but reported no indication of a tornado, which apparently formed directly over the field. Senator Lyndon B. Johnson's Senate Preparedness Subcommittee investigated the incident and concluded that the Eighth Air Force division commander and officers on duty at the time of the storm had taken adequate precautions in tying down the bombers with three-eighth inch cables, which the tornado had snapped like twine. The committee also noted that the storm "could not have been forecast in its full intensity in time to have averted the major damage which occurred."[44] Although the storm took no lives, the loss of much of the country's bombing capability at the height of the Cold War may have given more impetus to improving tornado forecasting than did the death of more than two hundred citizens a few months earlier.[45]

The air force awarded Fawbush and Miller Commendation medals and citations in January 1953. Secretary of the Air Force Thomas K. Finletter signed the citations for severe storm warning service the two officers rendered from March 1948 to April 1952. Each citation read in part,

> His technical ability and devotion to duty in developing a tornado and severe storm warning forecasting method had resulted in the saving of many lives and untold amounts of government property. Through the use of his forecasts, military installations throughout the United States have been enabled to prepare for tornadoes, high winds and hail, while pilots have been forewarned of icing conditions and degree of turbulence. These services have forestalled equipment damage amounting to millions of dollars. In cooperating with the United States Weather Bureau, he has also rendered a great service to the American public, by dispelling, to a large degree, the surprise factor inherent in most violent storms. By pioneering this new meteorological field, by helping to save human lives and great amounts of government property, and by giving unstintingly of his time to the development of severe weather forecasting methods, he has brought great credit upon himself and the United States Air Force.[46]

Both the tornado forecasting and warning systems were born during the 1940s. The needs of the military in times of war had prompted the Weather Bureau to create a severe storm warning network to notify military bases and munitions plants of approaching severe storms and tornadoes. But warning of an approaching tornado and predicting the appearance of a tornado at some future time are not the same thing. The Weather Bureau chose not to forecast tornadoes during this period because it did not believe its meteorologists had adequate knowledge or data to issue tornado forecasts with any reliability. One wonders whether Reichelderfer ever took into account the possibility that even an imperfect forecasting system could reduce tornado fatalities in the United States.

Although by 1952 the Fawbush-Miller technique for forecasting tornadoes had been quite successful, the Weather Bureau did not believe tornado forecasting for the general public was feasible. Reichelderfer feared the public would panic if the bureau issued a tornado forecast for a specific location, but an even greater fear was failure. In a time of postwar budget cuts, Reichelderfer was apprehensive about undertaking a new service that he did not believe could succeed. The bureau placed its emphasis on warning rather than forecasting, but when the civilian population learned that the air force had been successfully forecasting tornadoes, it began to apply pressure for parity. Only then did the Weather Bureau respond to the public's demands and institute a tornado forecasting service.

CHAPTER 4

A Civilian Severe Weather Forecasting Center

The Weather Bureau suffered a major public relations dilemma in the spring of 1952. The basis of the problem was the public's perception that the bureau could not and would not issue tornado forecasts to the civilian population, whereas law restricted the air force to issuing predictions for military establishments only. Both assumptions were false. Weather Bureau chief Reichelderfer had summarized his organization's position on tornado forecasting in a July 12, 1950, circular letter to all Weather Bureau stations. Evidently some offices had issued statements such as, "The Weather Bureau does not make tornado forecasts" or "We are not permitted to issue tornado forecasts." The chief urged caution in the use of tornado forecasts but explained to his employees that the bureau had the responsibility for warning the public of approaching destructive storms, and the district or local forecaster could, at his discretion, use the word *tornado* in the forecast or warning. Reichelderfer closed with the official viewpoint: "There is no regulation or order against the forecasting of tornadoes. Whenever the forecaster has a sound basis for predicting tornadoes, the forecast should include the prediction in as definite terms as the circumstances justify."[1]

As the 1952 tornado season approached, tension between the Air Weather Service and the Weather Bureau increased. Unlike the U.S. Air Force, the Weather Bureau had no centralized severe storm forecasting center comparable to the SWWC but gave the responsibility for issuing tornado forecasts to the meteorologists in the local Weather Bureau Offices, who rarely, if ever, issued tornado forecasts. Meanwhile, the SWWC issued forecasts throughout 1951 and early 1952, which the press often "leaked" to the public. After tornadoes struck parts of Alabama and Georgia on March 3, 1952, a United Press article asked why the area Weather Bureau Offices, armed with the same information and the SWWC forecast, declined to advise the public of the approaching severe weather. Ivan R. Tannehill, the Weather Bureau's forecast chief, gave the reason: "It is impossible to pinpoint a tornado by forecast." The article continued, "The weather bureau's position is that until it can pinpoint them with reasonable accuracy, it will lay off forecasting tornadoes." On the other hand, the air force, which did have tornado forecasting capability, repeated that it could not warn the public.[2] In response to this article Reichelderfer wrote Roy Calvin in the Washington United Press office that "there is no such law and the Weather Bureau would be opposed to legislation of this kind if it were proposed." What did exist was a working agreement between the bureau and the AWS that only one agency should issue forecasts to the public to avoid the chance of duplicating forecasts or giving conflicting advice.[3]

Throughout March and April the Oklahoma media, attempting to force someone to provide adequate tornado forecasting to the state, mounted a pro-SWWC and anti–Weather Bureau public relations campaign. Ken Miller of Tulsa radio station KVOO conducted a live interview with Fawbush and Miller on March 3. In his summary of the men's forecasting capability, the interviewer credited the SWWC with a 90 percent success rate, and neither officer corrected the statistic. The broadcast ended with a comment on the Weather Bureau's inability to provide the

SWWC's forecasts to the public: "Government officials have not yet untangled the red tape which would make this tornado forecasting service available to states, cities, communities and civilian population."[4] Throughout the period KVOO randomly aired derogatory statements and accused the Weather Bureau of not releasing SWWC's forecasts because of jealousy.[5]

Oklahoma City radio station KOMA, whose program manager Bob Eastman was a personal friend of SWWC forecaster Robert Miller, periodically criticized the Weather Bureau. On March 25, 1952, the station bragged that it had been the source of leaks to the United Press when the air force issued tornado forecasts for its installations. The same broadcast reported that two Weather Bureau men from Washington were in Oklahoma City "to place a curb on the release of Tinker forecasts." Reichelderfer had dispatched J. R. Fulks and Richard Schmidt to Tinker Field to review the status of tornado forecasting and to confer on the verification question, but the two had no orders or authority to restrict SWWC operations. Reichelderfer attributed the station's misrepresentation of the Weather Bureau to "a deliberate plan on the part of the Tinker Field's staff to get personal publicity and to force the Weather Bureau to release their forecast and not go through Kansas City."[6]

A front-page story in the March 28, 1952, *Daily Oklahoman* announced a conference chaired by Colonel Thomas S. Moorman, deputy chief of the AWS, where representatives of the Weather Bureau, newspapers, radio stations, press services, and the 2059th weather wing at Tinker Field would formulate plans for warning Oklahoma communities of approaching severe storms and tornadoes. The plan would allow the Oklahoma City Weather Bureau Office to release SWWC tornado forecasts directly to the media and avoid the delay of sending forecasts to Washington or Kansas City for checking or revision. Much of the article, and a companion piece that had appeared two days earlier, praised Fawbush and Miller for their high degree of accuracy.

What was the basis of the contention between the AWS and the Weather Bureau? The air force believed Fawbush, Miller, and the entire Tinker Field severe weather forecasting team had not received adequate public praise or publicity and resented the Weather Bureau questioning the accuracy of its verification statistics. In addition, the AWS believed the time had come for the Weather Bureau to issue its own tornado forecasts and create a centralized severe weather warning service that would assume responsibility for all civilian and military severe weather forecasting.[7] Reichelderfer recognized that the latter was at the bottom of the difference of opinion.[8]

The Weather Bureau felt inadequate to fight the air force or the public because of lack of public relations funds and White House restrictions on publically airing its differences with the AWS. From the bureau's viewpoint the greatest source of contention between the two organizations was the SWWC's misrepresentation of its forecasting accuracy. Press releases had left the impression in the public's mind that SWWC meteorologists could forecast individual tornadoes with greater than 90 percent accuracy and were predicting the time and place of occurrence of individual tornadoes.[9] J. R. Lloyd of the Bureau's Kansas City Regional Office and William Maughan from the Bureau's Oklahoma City Office had studied Tinker Field forecasts and compiled their own statistics. While the SWWC claimed a 92 percent accuracy rate for 1951, bureau meteorologists calculated only a 50 percent verification rate.[10] Discrepancies in success rates varied between the two organizations because they used different criteria for evaluation. According to Weather Bureau standards, a tornado had to occur within the prescribed area during the forecast time for a tornado forecast to be a success or verified. If a tornado occurred outside the forecast area or time frame, or if a tornado failed to occur at all, the forecast was a failure. The AWS counted its forecast a success or verified even if a tornado occurred within a severe thunderstorm forecast area or within 150 miles of a tornado forecast area. Reichelderfer knew the bureau could not publicize the discrepancies without implying a controversy with the air force, but he did point out to the AWS the unfairness of its presenting to the public a greater than 90 percent validity rate.[11] In an April 18, 1952, letter to Colonel Harold Smith, Tinker's Air Weather Wing commander, Reichelderfer, criticized the air force's indifference toward the Weather Bureau's position and urged the colonel to issue a brief statement to set press representatives straight on the statistics question. Reichelderfer insisted that the bureau would not have stood by and allowed unjust criticism of the AWS arising from Weather Bureau activities without an effort to bring out the facts in fairness to all concerned.[12]

The bureau did not believe the Tinker Field forecasting method could be applied to tornado forecasts for the general public because the SWWC revised its forecasts so frequently. Radio and press channels would have difficulty relaying altered forecasts to the public. Jokingly, Reichelderfer asked an air force representative how a commanding general of an air force base acted with so many tornado warnings being issued and then changed every few hours. "What does he do—send his planes away one hour and then call them back and then start them out again?"[13] The bureau also declined to issue SWWC's predictions because forecasts designed for the general public had to cover more areas for longer periods of time than the Tinker Field forecasts.

The Oklahoma congressional delegation, especially Senators Mike Monroney and Robert Kerr, wanted adequate tornado warnings for its citizens. The senators, staunch supporters of the SWWC, wanted the Oklahoma City Weather Bureau Office to cooperate with the Tinker group and become the center of a combined severe weather forecasting center. They feared dissension with the AWS might cause the Weather Bureau to transfer any tornado forecasting program it developed to its Kansas City Office.

If the public wanted tornado forecasts, the AWS wanted recognition, and the Weather Bureau wanted statistical truth and time to develop its own program, what did the media want? Reichelderfer thought the press wanted scandal. He related to four regional Weather Bureau directors on March 26 that "It's the old story that under present circumstances the entire press and much of the American public is looking for some scandal in the government and the minute they find something which they think may lead to information they don't have, why they ride it to death and we are almost at that point in this situation."[14]

From March through May 1952, Reichelderfer corresponded with many of the Bureau's severe storm experts, regional forecasters, and project leaders to criticize them for past inaction and to instill in them a sense of urgency. A March 25 memo for a project leader's conference best exemplified his frustration at the bureau's failure to create an adequate tornado forecasting system in the past. He maintained that it should have been obvious years ago that the public would expect continued improvement of warnings of severe storms that could cause loss of lives and property and would not accept a "can't-do-anything" explanation for failure to discover some means of advance notice. Forecasters should have given more concentrated and constant attention to the problem of tornado forecasting. The chief blamed his subordinates for the plight the bureau found itself in with the AWS. He believed a little foresight, or a little more attention by responsible project leaders to the suggestions and directives his office initiated, would have placed the Weather Bureau in the lead instead of in the embarrassing position of having to adopt something the air force developed.[15]

Of course, Reichelderfer could never allow the air force or the public to hear such words of criticism of his own organization. Neither could he afford to conduct a public battle with the air force over tornado forecasting and verification. Either action might invite a congressional investigation. The chief knew that the Interior Department had "suffered a 25 percent cut in appropriations because they got into a political mess of this kind," and he feared the bureau might suffer the same fate if Congress got involved. "If this thing should get into a Congressional investigation, it would simply boil down to the fact that the Tinker boys have been forecasting tornadoes and the Weather Bureau has not, and that would be pretty sad in this election year."[16] Reichelderfer reminded the project leaders that "the propaganda against the Weather bureau transmitted to members of congress by press and radio interests in Oklahoma are adversely influencing (and without real justification) the attitude of members of Congress toward the current appropriations estimates of the Bureau and may result in reductions which would impose curtailment in essential personnel and facilities in the national weather service."[17]

Reichelderfer's mistrust of the media and wariness of a congressional investigation may have been justified. The early 1950s was an era of unrest and fear in Washington. Wisconsin senator Joseph McCarthy led a hunt for subversives and communists throughout the government. No department, even the U.S. Army, was free from the senator's investigations. Chief Reichelderfer probably believed that the best way to escape scrutiny was not to attract attention, which a public fight with the air force would obviously do.

The Weather Bureau had at least one staunch supporter in Congress. Congressman Toby Morris of Lawton, Oklahoma, addressed the House on March 24, 1952, to request additional funding for Weather Bureau severe weather projects. As a representative of the people in one of the most tornado-prone states, Morris believed something could be done to protect not only his own constituents but all Americans from disasters such as the March 21 tornado outbreak. Public Law 657, approved by the Eightieth Congress on June 16, 1948, had authorized the chief of the Weather Bureau "to study fully and thoroughly the internal structure of thunderstorms, hurricanes, cyclones, and other severe atmospheric disturbances," but the bureau lacked money to carry out the directive. On April 4, 1952, Morris told the House that with adequate funding the bureau could employ radar and upper air soundings to make scientific investigations of severe storms and find answers that would save lives and property. The program would cost $1.5 million: $50,000 for severe local warning networks, $90,000 for radar networks, $75,000 for specialized studies, $900,000 for upper air observation equipment, $335,000 for purchase and maintenance of additional equipment, and $50,000 for forecasting for the public. The House rejected Morris's appeal for these additional funds, arguing that the original $27 million appropriation was adequate to meet the bureau's needs.[18]

In the spring of 1952 the Weather Bureau felt besieged from every direction—the air force, the public, the media, and Congress. Although Reichelderfer did not believe his meteorologists were adequately prepared to face the responsibility, he knew the Bureau had no choice but to do what the pressure groups wanted—issue tornado forecasts.

In the midst of the controversy the Weather Bureau quietly began issuing tornado forecasts. A group of research forecasters and supervising analysts at the Weather Bureau-Air Force-Navy Analysis Center (WBAN)[19] in Washington, D.C., had prepared tornado forecasts for March 9 and 10, but after considerable debate they decided their release was unwarranted. During the same period the SWWC had issued four forecasts but verified none.[20] After tornadoes struck near Wichita Falls and Graford, Texas, during the early evening of March 17,[21] WBAN issued its first tornado forecast, effective from 10 P.M. to 6 A.M. the following morning. The forecast simply said, "There is a possibility of tornadoes in eastern Texas and extreme southeastern Oklahoma tonight, spreading into southern Arkansas and Louisiana before daybreak." No tornadoes formed within the watch area, but fortunately very few people were aware of the forecast's existence because of the late hour. The bureau could not afford any more public relations debacles. By comparison, the air force's success rate for the same day was only 50 percent. The SWWC issued two tornado forecasts on March 17. The first, valid from 4 to 8 P.M. CST, was for an area within a fifty-mile radius of Guthrie, Texas. The second, for an area along and seventy-five miles south of a line from Wichita Falls to Denton, Texas, was in effect from 7 to 11 P.M. The two tornadoes that occurred were in the second forecast area.[22]

WBAN's next attempt to forecast tornadoes came just four days later, when atmospheric conditions over the southern plains indicated imminent severe weather. The Washington group issued tornado forecast number two at 11:55 A.M. CST on March 21, 1952, for virtually the same area the previous forecast had covered.[23] The official forecast simply said, "A few isolated tornadoes may occur late this afternoon and this evening in southern Arkansas, northern Louisiana, extreme southeastern Oklahoma, and the north half of eastern Texas." This time some of the public may have been aware of the forecast. The front page of the *Arkansas Democrat* contained a notice that "numerous thunderstorms and possible isolated tornadoes were forecast for early tonight in southern Arkansas by the U.S. Weather Bureau in Washington, D.C."[24] At 9:30 P.M. CST the bureau extended its forecast area to include central and western Kentucky, western Tennessee, and extreme southern Indiana until daybreak. Though they had the responsibility and authority to do so, no local Weather Bureau Office in the area issued its own tornado forecast for March 21.[25] During the afternoon and night seventeen tornadoes killed over two hundred citizens of Arkansas, Tennessee, Missouri, and Mississippi.[26]

Although the March 21 tornadoes killed so many Americans, meteorologists judged the forecasts a success because several of the tornadoes did occur within the watch areas; however, most of the hardest-hit areas were never included in the tornado forecast. At least ninety of Arkansas's 111 deaths, as well as the seventeen deaths in Missouri and the nine deaths in Mississippi, occurred outside of forecast areas.[27] The press took note of the bureau's failure. A United Press article in the March 23 *Memphis Commercial Appeal* complained that tornado victims had received little or no advance warning of the Friday devastation, although an air force storm warning expert said his staff had predicted a major part of the tornado activity. The article explained that the SWWC had issued confidential warnings to military bases in the ravaged area hours before civilians had any word of possible disaster. Fawbush reiterated the air force's position that law prohibited the SWWC from releasing any of the several forecasts it had issued that day to the public. The first, in effect from 3 to 8 P.M. CST, was for an area along a line from fifty miles north of Greenville, Texas, to Monticello, Arkansas. The revised forecast in effect after 4:20 P.M. CST covered the area bounded by Walnut Ridge, Arkansas, Dyersburg, Tennessee, Monticello, Arkansas, Shreveport, Louisiana, and Texarkana, Texas. The deadliest tornadoes struck Arkansas during this time. Fawbush and Miller emphasized that although they had a reputation for being able to "pinpoint" tornadoes in advance, they were not in competition with civilian forecasters, even though it appeared that way.[28] The press failed to consider that unlike the military's communications system for spreading word of a tornado forecast, a comparable method for warning the public did not exist. Cooperation among the news media, law enforcement agencies, and the bureau was voluntary, and while military bases had prescribed procedures to follow to reduce loss of property and life, the public was woefully unaware of what actions to take when the bureau issued a tornado forecast.

In spite of the bureau's failure to forewarn many citizens of the potential for hazardous weather, some of the press offered favorable comments. An Associated Press article in the March 23 *Arkansas Gazette* said, "The Weather Bureau isn't claiming any special credit, but it did a pretty good job of forecasting the tornadoes which devastated areas of the South Friday night."[29] A similar article in the March 22 *Commercial Appeal* explained that although "tornadoes have long been considered unpredictable, the Weather Bureau in Washington, in an unusual forecast, called the turn on the rash of twisters in the Mid-South yesterday. Within two hours after a warning forecast had moved over the news wires, the first tornado struck Dierks, Arkansas."[30]

The Weather Bureau issued fifty-seven tornado forecasts in 1952 but verified only eleven, for a 19 percent success rate.[31] Fortunately, after the March 21 outbreak, tornadoes killed only eleven Americans during the remainder of the year. As the year progressed, publicity died down. One exception was the April 11 forecast WBAN issued for parts of Oklahoma. The next day the *Daily Oklahoman* called the state tornado alert "nothing but hot air" and criticized the Kansas City Weather Bureau Office for not "dusting off its crystal ball." The article indicated that a conflict had existed between the Oklahoma City Office, which had consulted SWWC, and the Kansas City forecasters. When Tinker Field meteorologists reported Oklahoma would have no severe weather for several days, the Weather Bureau canceled the forecast.[32]

WBAN issued experimental tornado forecasts from March 17 to May 21, 1952, when the bureau established a separate Severe Weather Unit (SWU)[33] at the Analysis Center to issue both tornado and severe thunderstorm bulletins as needed on a twenty-four-hour basis. During the winter of 1952–1953 bureau personnel discussed ways to improve severe weather forecasting in the upcoming tornado season. Although the SWU had issued several tornado forecasts in 1952, the ultimate forecast responsibility remained with the district office. Unfortunately, the preferred severe weather forecasting techniques required extensive data plotting and analysis, which district forecasters had no time to perform. To solve this problem the Kansas City district office instituted a program on March 1, 1953, to determine what existing personnel could do to aid district forecasters. The office designated a severe weather forecaster, Donald S. Foster, who could adjust his schedule to correspond to times of maximum tornado activity. Foster concentrated on analyzing data for the Kansas

City Office's jurisdiction (Kansas, Missouri, Oklahoma, and Nebraska) and areas likely to be affected by tornadoes. For future reference he maintained a severe weather log, which included copies of all local severe weather forecasts, bulletins from Washington and Tinker Field, and severe weather reports. In addition, Foster placed charts for days when tornadoes occurred or the Kansas City Office issued forecasts in a tornado case file for future study. At the end of August he calculated the program's success. Of the 150 tornadoes that struck within the district's area of responsibility from March 1 through August 31, 74 percent occurred within a severe weather forecast's time period and area. In comparison, the Weather Bureau's severe weather unit in Washington had a 23 percent verification rate for 1953. To the Weather Bureau establishment, this experiment emphasized the need for more personnel at the district level and more severe weather forecasting training.[34]

Five forecasters from the central office and field stations composed the SWU temporarily until a permanent staff could receive adequate severe weather training. To attract young field forecasters to the fledgling unit, the bureau's Scientific Services Division offered a series of two-week training courses on the formation of severe weather, AWS's methods of tornado forecasting, and the latest tornado research. The first five members, Joseph Galway, James Carr, Robert Martin, Allen Brunstein, and David Stowell, were relatively young college-educated men who had received most of their meteorological training in the military. According to Galway, the group was "intentionally selected because they were less likely to have preconceived ideas in the area of severe storm forecasting." Kenneth Barnett, a Navy-trained meteorologist, became SWU supervisor in December 1952.[35]

Meteorologists at the Chicago and Kansas City Regional Forecast Offices developed worksheets or lists of parameters that favored tornado formation for SWU's use. Local offices teletyped temperature, wind velocity and direction, barometric pressure, relative humidity, dew point, and sky conditions to the Washington center, where forecasters combined the data with readings from twice-daily upper-air soundings and plotted the results on a single map or composite chart. When enough criteria were present to warrant a possible tornado forecast, SWU consulted the forecast offices within the threatened area to get a better idea of the actual meteorological situation. The final decision to issue the forecast lay with the local office.

SWU's job was to issue clear, concise severe weather bulletins (forecasts) whenever needed to cover all severe local conditions, including thunderstorms, hail, and tornadoes. Designation of the forecast areas was free-form. Joseph Galway recalled that in the early days the areas were not necessarily boxes but, frequently, circles.[36] A few years later forecasters used plastic templates with scales from ten to seventy miles to draw watch boxes on maps. Regardless of the forecast area's shape, the written forecast had to follow precise guidelines. A bulletin could cover only one contiguous area for a designated period of time—usually six hours—had to be complete within itself, and could not refer to previous ones. The forecaster typed the bulletin on paper and passed the bulletin to the communicator in the communications room, who made a tape and forwarded the message by teletype to the local offices involved.[37] Because they went only to Weather Bureau offices, the early forecasts used extensive meteorological contractions and abbreviations, such as in this typical bulletin:

> SVR WX BLTN 011830Z
> WBC BLTN NR 1. COLD FNT 50S MKC 30N ICT 30E GAG
> TO NEAR CDS AND LBB AT 1830Z WL MOV EWD AND SWD
> ABT 10 KTS ACPYD AFT 2000Z BY SCTD TSTMS BCMG BY
> 0000Z NEARLY SOLID LINE TSTMS ALG TEX PNHDL AND
> OKLA PTN OF FNT. THESE TSTMS ACPYD BY MDTLY STG
> SFC GUSTS AND SVR TURBC ALL LVLS. SCTD SHWR CNDS
> WITH CSDRBL TURBC ALL LVLS CNTRL OKLA AHD COLD FNT.

SWU teletyped the bulletins to New York City, Miami, Kansas City, and Salt Lake City for retransmission to local offices.[38]

Reichelderfer changed the name of the Severe Weather Unit to the Severe Local Storm Warning Center (SELS) on June 17, 1953, and directed Barnett to reduce both the size and frequency of forecast areas. During this period the air force issued tornado forecasts for only twelve to fifteen thousand square miles, whereas the Weather Bureau's watch areas covered thirty to forty thousand square miles and sometimes entire states. To aid in this project and to attempt to bring some prestige to the fledgling severe weather unit, Barnett requested a research forecaster for SELS in the summer of 1953. Robert G. Beebe, who had directed tornado research at the Atlanta office, joined SELS in October. Before Beebe's arrival the SELS forecasters had a dismal publication record; only one paper had appeared in a scholarly journal. From 1953 to 1957 the record did not improve substantially. Journals published only twenty-three of eighty-seven papers the SELS meteorologists produced. Many of these appeared in the "Correspondence to the Editor" section of the *Bulletin of the American Meteorological Society* because the bureau's central office had to approve a paper's submission to a meteorological journal, but it had no authority over the correspondence section. Galway believed that although some of the papers may not have been worthy of publication, those in positions of authority within the meteorological community blocked publication of some credible papers because of dissatisfaction with the bureau's foray into tornado forecasting.[39]

Nature was unkind to the Weather Bureau's fledgling SWU in 1953, the first full year of its existence. Before the year ended 412 reported tornadoes would claim 515 American lives, including a total of 323 at just three sites, Waco, Flint, and Worcester. The year was the fifth worst ever for tornado deaths in the United States and by far the worst since the institution of any type of tornado forecasting or warning system.[40]

Texas's greatest tornado disaster occurred in Waco on May 11,

1953. After consulting with Washington, the New Orleans Office issued a tornado forecast at 9:30 A.M. for an area of Texas bounded by San Angelo, Waco, Wichita Falls, and Big Spring.[41] Although a tornado had killed thirteen in the San Angelo area in the early afternoon, C. A. Anderson of the Waco Weather Bureau Office, not wishing to alarm citizens unnecessarily, told the *Waco Times Herald* that there was no cause for worry in central Texas. No tornadoes had developed in the area yet, and any that did would strike west of Waco. Instead, Anderson predicted "strong winds, scattered thunderstorms, and probably some hail in limited areas."[42] Besides, all Waco residents knew that according to Huaco Indian legend, a rim of hills surrounding their town on the banks of the Brazos River ensured its protection from tornadoes. Ironically, Anderson had told the *Waco Times Herald* only one year earlier that he wanted to set up a tornado warning service for Waco and Central Texas, and although he was not expecting any tornadoes, "if some of the twisters should show up any time, a report to his office would help warn people ahead of it. The Waco area was not immune from the storms."[43] Some citizens were aware of the tornado forecast, but most went about their daily routine, paying little attention to the weather. Roger Conger, an employee at a Waco gas station, was on duty when his wife telephoned him to ask whether he had heard the tornado alert on the radio. His attitude was the common one of the day: he laughed and told her he had not heard the forecast, but it would not make any difference because Waco had never had a tornado.[44] Many residents, especially those who did not own a television or had not been listening to the radio, were unaware of the tornado forecast or the fact that a tornado had touched down at Hewitt, about eight miles southwest of Waco. The city had no warning system. A massive wedge-shaped funnel cloud dropped out of the clouds about 4:30 P.M. and roared through residential areas on its way to the heart of the business district. When it lifted, 114 of Waco's 85,000 citizens were dead. The F5 tornado had destroyed much of downtown Waco.[45]

Two other metropolitan areas suffered similarly destructive tornadoes in 1953 when, in June, one of the century's most devastating weather systems moved from the Plains through the upper Midwest to New England. Working around-the-clock shifts, the SWU in Washington issued tornado forecasts for large areas of Michigan and Ohio on June 8. In spite of the forewarning 141 died, including 115 in Flint, Michigan, where an F5 tornado plowed a half-mile strip through a residential area of the city of about two hundred thousand.[46] This tornado has the distinction of being the last single tornado in the United States to kill more than one hundred people. The third disaster struck the next day. No one had informed the two hundred thousand residents of Worcester, Massachusetts, of the possibility of tornadoes in their area. The Buffalo, New York, Weather Bureau Office had predicted unsettled conditions and had warned the residents of western New York that a tornado might occur in their vicinity, but the Boston office, which had responsibility for Worcester, had predicted warm temperatures and the likelihood of an afternoon severe thunderstorm. At 3:25 P.M. a massive tornado tore through the unsuspecting city, killing ninety-four and injuring more than twelve hundred residents. On June 10 Reichelderfer wrote SWU director A. K. Showalter, "It is tragic that the disastrous Worcester tornado occurred so soon after the Flint, Michigan, tornado and that our Severe Storm Warning service was so close to issuance of warnings in each case but actually missed!" A year later the bureau chief explained to a House Committee on Appropriations hearing that the SWU had forecast the Waco, San Angelo, Cleveland, and Flint tornadoes[47] but had missed the Worcester one partly because the upper-air observations in that area were not sufficient and partly because tornadoes were so rare in Massachusetts the forecasters could hardly believe it would happen.[48]

Public and media response to tornado forecasting varied from one locale to another. Often individuals griped to their local Weather Bureau Offices. The Indianapolis office reported

TABLE 5

Tornado Statistics for the 1950s

Year	Number of Tornadoes Reported	Tornado Deaths	Most Deaths in a Single Tornado
1950	200	70	18
1951	262	34	6
1952	240	229	57
1953	421	515	115
1954	550	36	6
1955	593	126	80
1956	504	83	25
1957	856	192	44
1958	564	66	19
1959	604	58	21

SOURCE: *Storm Data 1995*, 40.

to Reichelderfer that one of the city's shopkeepers had complained that a tornado forecast had cost her business when shoppers fled her store in panic after they heard a weather bulletin on the radio.[49] Another resident described the panic among mothers and children at a school rehearsal during a tornado forecast and urged the bureau not to frighten the people to the point where they could not respond in the event of an actual tornado.[50] An *Atlanta Journal* editorial on March 25, 1952, described the panic in sections of the South when the bureau issued one of the first forecasts. Some residents fled to their cellars, while others flocked to churches to ask for divine intervention. "There has been nothing like it since the radio announced an invasion from Mars many years ago and threw a section of the nation into hysteria."[51] A May 11, 1952, *Commercial Appeal* editorial suggested that because there was nothing an average citizen could do to escape a tornado except worry, "a lot of people would prefer not to be told that they may be blown away." The editorial questioned the value of "creating apprehension among those all but powerless to do anything except quake."

These editorials addressed valid points. A lack of shelter was a common problem in tornado-prone areas. In many parts of the country, especially the Southeast and Texas, most houses lacked basements, and though storm cellars were common in rural areas, they did not exist in cities. Many of those who were aware of a tornado forecast had no place to take shelter, and the evacuation of urban areas was impractical. The media's misuse of the terminology may have contributed to these citizen's fears of tornado forecasts. The Weather Bureau issued a *bulletin*, which was its word for *forecast*. A tornado *forecast* meant that a possibility existed for tornadoes to form somewhere in the designated area. It did not mean that a tornado had already formed and was on the ground. A tornado *warning*, however, meant that someone had spotted a tornado and that it was approaching a specific area. A forecast warranted precaution, but a warning required immediate action. Unfortunately for the public, the media and the Weather Bureau frequently used *bulletin*, *forecast*, and *warning* interchangeably. To clarify the misunderstandings and attempt to alleviate some of the public's fear that was associated with the term *bulletin*, the bureau replaced the word *bulletin* with the word *forecast* in 1954. In 1966 the bureau instituted the use of *watch* for all tornado and severe thunderstorm forecasts. The meaning of *warning* remained the same throughout the time.

Not all media coverage was negative. The *Joplin Globe* warned Missouri citizens not to disregard the forecasts because they were becoming increasingly accurate. "If the forecasts are correct only a small percentage of the time and thereby save life and suffering, it will be a tremendous public service. Better not hoot at the tornado forecasts. The next one may prove correct."[52] The *Cincinnati Times-Star* called the forecasts "extremely valuable," and the *Enid* (Oklahoma) *Daily Eagle* said such warnings were well worth the effort if they saved but one life.[53] The *Daily*

Oklahoman published a letter from an Enid resident who thanked the bureau for recent warnings and asked that they be continued because the average person could not foresee severe weather conditions, especially at night.[54]

Barnett failed to fulfill Reichelderfer's mandate to reduce the size and number of forecasts. Increasing criticism for this failure, along with the mounting tide of complaints about forecasting inadequacies, such as that for the Worcester tornado, led to Barnett's leaving the bureau for the Army Signal Corps. Donald House, a district forecaster at the Kansas City Office, became the new SELS supervisor in March 1954.[55]

Significant changes in forecasting procedure occurred in 1954. Although the district forecast office had the final say on issuing forecasts for its jurisdiction, SELS began making the final decision if a tornado forecast involved more than one district office. Not until 1958 did SELS assume sole authority for the issuance of severe weather forecasts throughout the United States. House was able to fulfill Reichelderfer's mandate to reduce the forecast area size to 16,320 square miles, less than half of the previous year's area.[56] Average lead times for the forecasts ranged from fifteen to fifty-eight minutes (table 6).

During tornado forecasting's developmental stage, some meteorologists, believing that one method did not apply universally, devised forecasting methods unique to their state or region. F. G. Shuman and L. P. Carstensen proposed that tornadoes would form in Kansas and Nebraska within twelve hours when tropical air was over or southeast of the area, a deep cold air mass was west of the area, a well-defined pressure trough at 700 millibars lay above the area, and the temperature trough and pressure trough were out of phase.[57] Others produced similar parameters for the Gulf states, Ohio, Kentucky, Tennessee, and the Mississippi Valley. These methods shared one fatal flaw: The forecast area's size and time period were too large.[58]

Rumors that SELS would relocate in Kansas City surfaced during the spring of 1954. The press in the Midwest and the

TABLE 6

Summary of Verifications

Year	No. of Forecasts	Verified	% Verified	Lead Time (in minutes)	Average Area (in sq. miles)
1952	57	11	19	58	35,980
1953	176	40	23	20	34,200
1954	236	51	22	15	16,320
1955	211	70	33	31	21,800
1956	188	102	54	47	30,000
1957	282	172	61	31	41,930
1958	116	39	34	28	22,860
1959	130	56	43	50	25,770
Average	174	68	39	35	28,607

Source: Joseph Galway, Verification Statistics, 1966, mimeographed copy available at Storm Prediction Center, Norman, Oklahoma.

Plains states had been pressuring Chief Reichelderfer to move the Weather Bureau's severe storms operations from the East Coast, where tornadoes were relatively uncommon, to the midwest and plains states, areas most prone to tornado activity. Two major factors in Kansas City's favor were its central location between New Orleans and Chicago, the two other regional forecast offices most involved with tornadoes, and its role as a Weather Bureau teletype communications center. The bureau notified its offices of the official move on August 23, 1954.[59] The supervisor and three forecasters arrived at the new location the same week and began issuing tornado forecasts for areas west of the 90th meridian on September 1, while Washington continued forecasting for the eastern part of the country. When two SELS forecasters from Washington arrived in Kansas City two weeks later, the severe weather center transfer was complete. For the next forty years the Weather Bureau's severe weather forecasting office would remain in Kansas City, which ironically had been John Park Finley's home base for tornado observations more than seventy years earlier.

Several significant events occurred during SELS's first years in Kansas City. The Weather Bureau instituted a Radar Report and Warning Coordination (RAWARC) teletype system to send radar reports, warnings, and upper-air information to Kansas City on September 15, 1955. The direct teletype transmission of radar data to SELS necessitated the formation of a Radar Analysis and Development Unit (RADU) to collate and analyze the information and to transmit hourly summaries to the field offices. In addition, RADU freed SELS forecasters from plotting maps of radar reports.[60]

Cooperation between SELS and SWWC became a reality in January 1956, when the AWS moved its severe weather unit to Kansas City. The object was to share expertise and to avoid duplication, a mandate of the Eisenhower administration, but each unit operated as an independent entity issuing forecasts for its constituency. The air force had brought with it premier tornado forecaster Robert Miller's composite chart technique, which plotted location and orientation of various atmospheric features on a single map. The composite chart and its companion parameter checklist became the basis of severe weather forecasting as meteorologists essentially learned to recognize distinctive patterns that preceded the development of severe thunderstorms, hail, or tornadoes. The air force disbanded SWWC in 1961, and SELS assumed responsibility for all severe weather forecasting in the United States.[61]

The 1965 tornado season was a particularly severe one. On Palm Sunday, April 11, tornadoes devastated six midwestern states, causing 256 fatalities. Allen Pearson, chief of the Weather Bureau's Emergency Warnings Branch in Washington, D.C., led a survey of the impacted areas to determine what, if anything, the bureau could do in the future to avoid such a loss of life. In August 1965 House moved to the Environmental Science Services Administration (the Weather Bureau's parent organization from 1965 to 1970) headquarters, and Pearson, who had spent most of his career in the Pacific and "knew absolutely nothing about tornado forecasting" became the new head of SELS, renamed the National Severe Storms Forecast Center (NSSFC) in 1966. Pearson initially tried to emulate House's methods, but he soon realized that the analysis of data, especially radar reports, was so time consuming for lead forecasters that their results would be "mediocre at best." Dr. George Cressman, chief of the Weather Bureau and an early leader in computer modeling, had improved the computer capability of the Washington office, but in an era of congressional demands for reduction of expenditures, funds to provide computers for the two other major forecast centers, the National Hurricane Center and NSSFC, were lacking.

Pearson was able to give the NSSFC national exposure through yearly appearances on the NBC *Today Show* beginning in 1968. To promote the implementation of the SKYWARN program, a local-level severe storm spotting and warning program, the NSSFC director toured the country during the spring for five years, lecturing to groups in towns with National Weather Service offices in an effort to get countrywide preparedness. By this time Pearson had developed an "amplified SELS log," which permitted the preparation of tornado strike probabilities. Based on this technique, he issued a bulletin on April 2, 1974, notifying NWS offices in a large area east of the Mississippi River of a potential large tornado outbreak during the next two days. The April 3–4 outbreak was the worst in the country's recorded history, and for his work in preparing for the onslaught Pearson received the Department of Commerce's Gold Medal. Media exposure after the outbreak brought NSSFC's demands for more funds to Congress's attention and eventually led to the development and acquisition of better technology for severe weather forecasting.[62]

One significant addition to NSSFC under Pearson's administration was the Techniques Development Unit (TDU) to serve as an interface between the research and operational (forecasting) communities in 1976. Among TDU's activities were scientific studies on conditions that produced severe weather, development and implementation of computer technology, improvement of forecasting procedures, evaluation of data sources, and verification of the NWS's severe weather watch/warning program. The first TDU chief was Joseph Schaefer.[63]

When Allen Pearson retired from NSSFC in 1980, Frederick Ostby became the new director. He had joined NSSFC as deputy director in 1972. A native of Connecticut, where hurricanes and "northeasters" were common but tornadoes quite rare, Ostby in 1956 had joined the private weather forecasting organization, Travelers Weather Service, headquartered in Hartford. His primary duties were television and radio weathercasting. He and his colleagues at Travelers were among the first to receive the American Meteorological Society's Seal of Approval for both television and radio in 1960. During Ostby's tenure, which lasted until 1995, NSSFC shifted from paper and pencil to computers (the chapter on technological improvements details these changes). The computers not only provided the severe weather forecasters with more timely meteorological data and better forecasting tools but also freed them from the numerous clerical tasks, such as archiving severe weather reports. This, in turn, allowed the forecasters to issue new and more frequent forecast products, such as "Day-2 Outlook" and "mesoscale discussions," and to provide more detailed guidance for NWS field offices. In addition, the severe weather specialists could devote more time to individual research studies.[64] In 1994 NSSFC began posting its products on the Internet, and in 1995 it established World Wide Web pages that included not only severe weather forecasts but also severe weather archives. When Ostby retired from NSSFC in 1995, he noted that he was especially proud of NSSFC's improvement in anticipating severe weather outbreaks and reducing the death toll from tornadoes during the years he had served as director.[65]

Ostby's successor at NSSFC was former TDU chief Dr. Joseph Schaefer, who had extensive experience in severe weather forecasting and research. Schaefer's scientific interests focused on forecasting techniques, tornado climatology, forecast verification, and the application of technology to forecasting. Under Schaefer, NSSFC moved to Norman, Oklahoma, and received a new name: the Storm Prediction Center (SPC), a part of the National Centers for Environmental Prediction. The move was part of the National Weather Service's modernization plan instituted in the 1990s as a money-saving step, but the concept was not new; Allen Pearson had recommended the move in the 1970s.[66] Other factors besides the budget influenced the relocation: (1) Improvement in communications had negated the original need for locating the severe storms center in Kansas City; (2) the National Aviation Weather Advisory Unit, a part of NSSFC, had surpassed the severe weather unit in size; (3) a need existed for an All-Hazards Forecast Center, which included winter weather and excessive precipitation forecasting; and (4) the National Severe Storms Laboratory was already in Norman.[67]

In summary, although Miller and Fawbush at Tinker Field's SWWC had demonstrated the feasibility of forecasting tornadoes for a specific time and place by the early 1950s, Weather Bureau chief Reichelderfer hesitated to allow his organization to offer the same service to the general public. He believed that SWWC had not been as successful as it had reported and that the bureau could not adapt the air force's methods to civilian forecasting. Continued pressures from the media, Congress, and Oklahoma citizens forced a reluctant Weather Bureau to initiate a tornado forecasting service in March 1952. In spite of Reichelderfer's apprehensions that his forecasters were not prepared, the first bureau forecasters did an adequate job in a time when severe weather forecasting techniques were in their infant stages. To improve tornado and severe thunderstorm forecasting capability, the bureau created a separate severe weather unit at Washington, D.C. By the time the SELS relocated to Kansas City in 1954, tornado forecasts had become common. But the issuance of a tornado forecast was only a small part of the effort to save American lives and property. For the entire process to be effective, it needed three more ingredients: a system for spotting tornadoes, a method for warning the public of the storm's approach, and a strategy for educating the public. Along with the development of tornado forecasting, initial programs in these three areas would be in place by the end of the 1950s.

Tornadoes like this one near Dimmitt, Texas, are common in Tornado Alley. Courtesy of the National Severe Storms Laboratory.

A strong tornado at Henderson, Tennessee, in 1952 left behind demolished houses and toppled trees. Courtesy of Johnnie Cooper.

Lieutenant Colonel Ernest J. Fawbush (left) and Captain Robert E. Miller (right) plotting a tornado forecast in 1951. Courtesy of the U.S. Air Force.

Dr. Francis W. Reichelderfer, chief of the U.S. Weather Bureau, 1939–1963. Courtesy of NOAA.

Head chartist Andy Anderson plotting weather data on a map of the United States, 1956. Courtesy of the Storm Prediction Center.

H. Swenson (left), SELS director Donald House (center), and Joseph Galway (right) checking radar plots in 1956. Courtesy of the Storm Prediction Center.

A radar console, 1958. Courtesy of the Storm Prediction Center.

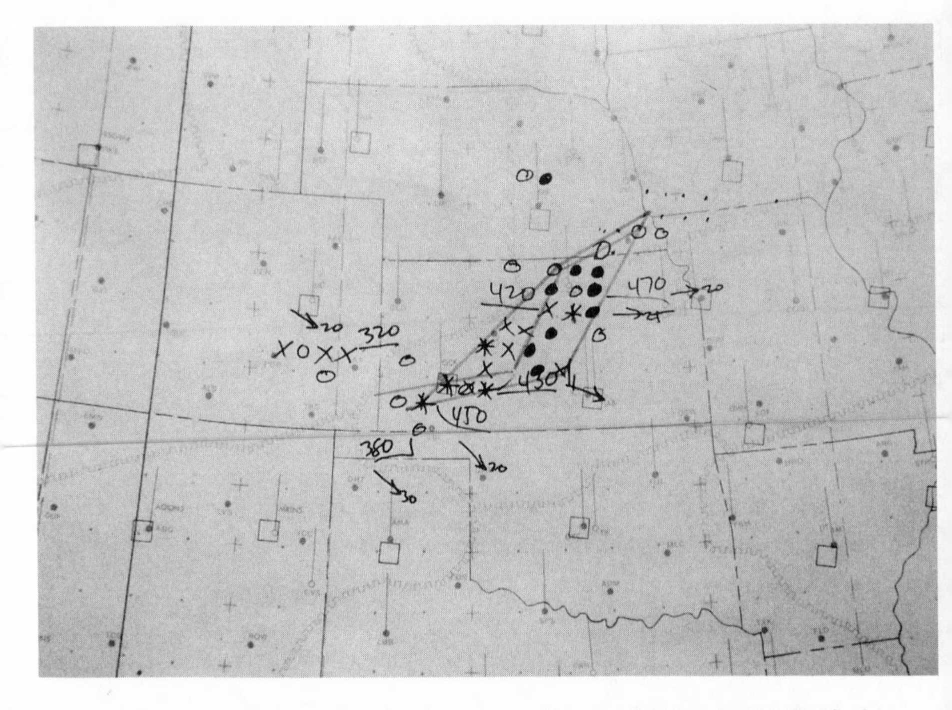

The Radar Analysis and Development Unit plotted radar data (intensity of echoes, cloud heights, and storm movement) on a large plastic-coated wall map. Courtesy of the Storm Prediction Center.

Allen Pearson, Director of the National Severe Storms Forecast Center, 1965–1980. Courtesy of Allen Pearson.

Teletype machines transmitted data from radar sites and local Weather Bureau offices to SELS in Kansas City. Courtesy of the Storm Prediction Center.

Before the computer age, SELS depended on paper maps and charts. Courtesy of the Storm Prediction Center.

TOTO (Totable Tornado Observatory), an instrument package designed to be placed in the path of a tornado to obtain meteorological data. Courtesy of the National Severe Storms Laboratory.

The Union City, Oklahoma, tornado of May 24, 1973, the first tornado captured by NSSL Doppler radar and an NSSL chase team. Courtesy of the National Severe Storms Laboratory.

The National Severe Storms Laboratory's first Doppler weather radar, located in Norman, Oklahoma, led to the National Weather Service's NEXRAD network. Courtesy of the National Severe Storms Laboratory.

Supercells like this one near Miami, Texas, produce many violent tornadoes. Courtesy of the National Severe Storms Laboratory.

A wall cloud may be a harbinger of a tornado. Courtesy of the National Severe Storms Laboratory.

National Severe Storms Laboratory/University of Oklahoma mobile radar trucks. Courtesy of the National Severe Storms Laboratory.

Dr. Joseph Schaefer, Director of the Storm Prediction Center. Courtesy of the Storm Prediction Center.

Storm Prediction Center lead forecaster Steve Weiss utilizes computers and paper charts to make severe weather forecasts. Courtesy of Steve Corfidi.

Storm Prediction Center mesoscale forecaster Jeff McBeath-Peters at an AWIPS three-headed workstation. Courtesy of Steve Corfidi.

CHAPTER 5

WARNING THE PUBLIC

Although the Weather Bureau and the AWS had a monopoly on tornado forecasting during the 1950s, no one entity had the responsibility for spotting tornadoes, warning people of their approach, or educating them on actions to take when a tornado was approaching. Universities and other scientific research facilities searched for ways to identify a tornado in its formative stages. Law enforcement agencies, the Civil Defense, community organizations, and the infant television industry shouldered the responsibility for warning civilians of an approaching tornado and for providing them with safety instructions.

As an aid in detecting and tracking the fleeting tornado, meteorologists employed radar, a relatively new technology. Radar, an acronym for *radio detection and ranging*, is a sounding device that transmits short bursts, or pulses, of radio waves and "listens" for their return. When the wave, which travels at the speed of light, encounters a scattering object such as a raindrop, the object scatters the wave's energy in all directions. Some of the energy is deflected toward the radar site, where an antenna detects it and feeds the energy (signal) to an indicator. The indicator converts the signal into a useable form, usually a video display on the fluorescent screen of a cathode-ray tube (a radar scope). Because distance is the product of speed and time, the radar determines the distance to a reflecting object by measuring the time needed for a signal to return to the radar site.

Almost everything, from airplanes to insects, reflects radar waves, and even areas of light rain are well defined by modern weather radars.[1]

The scientific basis of radar began with the work of Scottish physicist James Clerk Maxwell, who explored the relationship between electricity and magnetism and showed through mathematical equations that the two phenomena were indissolubly bound together. Maxwell proposed that visible light was only one member of a whole family of electromagnetic waves that traveled at the speed of light. German physicist Heinrich Hertz generated and detected electromagnetic waves a million times longer than visible light and showed that these radio waves had the same properties of visible light waves, including speed and the ability to be reflected. Guglielmo Marconi, an Italian physicist, thought the radio waves might be used for signaling. When the Italian government showed little interest in his work, Marconi went to England, where he succeeded in transmitting radio waves several miles without wires. He applied for a patent for the first radio in 1896, and shortly after the new century began he sent the first radio waves across the Atlantic.

In 1904 German engineer Christian Hülsmeyer demonstrated the telemobiloscope, an apparatus that used radio waves to allow a steamship captain to determine the distance and direction of another ship up to three miles away, but his work attracted little attention. During the next three decades scientists and engineers in the United States, Great Britain, and Germany experimented with pulse radio waves. With the threat of war on the horizon, Germany's main thrust was to develop a technique for detecting surface ships, and Great Britain concentrated on a method for detecting aircraft. By June 1935 Scottish physicist Robert Watson-Watt at the Radio Research Station of the British Department of Scientific and Industrial Research had developed a primitive pulse radar capable of detecting airplanes at ranges up to seventeen miles. Before World War II began the British had installed a chain of radar-warning stations along its shores, and the Germans had three types of radar, one for naval purposes, another for fighter plane guidance, and a third for spotting aircraft along the French and German coasts.[2]

The first radar sets operated at wavelengths measured in meters, but the long wavelengths had poor sensitivity to precipitation. John Randall and Henry Boot at the University of Birmingham developed the resonant magnetron, a device that generated microwaves, in 1940. One consequence of radar's shift to shorter wavelengths was improved ability to detect rain and snow. A ten-centimeter radar set tracked a thunderstorm with hail a distance of seven miles along the English coast on February 20, 1941, about the same time that Massachusetts Institute of Technology Radiation Laboratory scientists observed precipitation echoes on their radar scope. Meteorologists quickly began to recognize the characteristic radar echoes that widespread rain showers, frontal thunderstorms, and tropical storms produced.[3]

Army Signal Corps Laboratory engineers modified gun-laying or bombsight radar for use as weather radar. The navy gave the Weather Bureau twenty-five surplus airborne APS-2F radar sets in 1946, and immediately the bureau began experimental work on modifying the radar for storm detection. The fifty-kilowatt radar, designated AN/APS-2, operated on a frequency of 3,300 megahertz and had a range of slightly over one hundred miles. The bureau commissioned its first radar set at Washington, D.C., on February 14, 1947, and by the end of the year radar was operating at offices in Wichita, Kansas, Norfolk, Nebraska, and Wichita Falls, Texas.[4] The Raytheon Manufacturing Company of Waltham, Massachusetts, produced the three-centimeter AN/CPS-9, the first radar designed specifically for meteorologists, for the AWS in 1949. The Weather Bureau continued to use modified surplus military radars commissioned WSR-1, -2, and -3.[5] Radar installation at Weather Bureau stations was slow. By 1953 the radar network consisted of only twenty-three stations, fourteen of which belonged entirely to the

Weather Bureau. The remaining nine sites, which educational institutions, research associations, or governmental agencies owned, were cooperative efforts. Because modification funds came from its annual appropriations, the bureau could add only two or three radar sites to the system yearly.[6]

From the early days of radar, meteorologists recognized characteristic shapes of thunderstorms and hurricanes on the radar screen, but other hazardous conditions such as tornadoes, hail, high winds, and heavy rain did not appear to display distinctive radar echoes. While studying a thunderstorm on radar on April 9, 1953, Glenn E. Stout and his associates at the Illinois State Water Survey noticed an unusual echo approximately ten miles from the radar site at the University of Illinois Airport near Champaign. The echo "developed a tail, which as it moved eastward at about 48 mph, curled into a cyclonic whirlpool about 1.5 miles in diameter." When reports that a destructive tornado had followed the same path as the storm on the radar screen reached Stout, he realized that the echo with a tail (later termed a hook echo) had been a tornado.[7] Although pictures of the radar screen appeared in the June 1953 *Weatherwise*, more observations would be necessary before meteorologists could conclude that the hook echo always corresponded to a tornado. Further verification came that year when scientists at the Agricultural and Mechanical College of Texas (hereafter Texas A&M College)[8] analyzed photographs of the radar screen at the time of the Waco tornado, and meteorologists at MIT studied similar pictures taken during the Worcester twister. Both agreed that a tornado did display a unique echo on a radar screen.[9]

When Texas A&M student Don Moore looked at the screen of the modified SO-7N ten-centimeter cm radar on the afternoon of May 11, 1953, he noticed an unusual, quickly moving cloud formation. Moore worked part-time on a research project that studied the application of radar to weather forecasting. Part of his job was to take routine photographs of the echoes on the scope. The photo Moore took at 4:32 P.M. showed five large echoes, but neither Moore nor anyone on the meteorology staff paid particular attention to the comma-shaped echo. They were unaware that the Weather Bureau had posted a tornado alert for the area about eighty miles north of the College Station radar site. Within minutes the echo moved over the city of Waco. That evening, news that a tornado had devastated the city and killed 114 of its citizens spread throughout the state.[10]

When Texas A&M meteorologists studied the radar photographs, they realized the Waco tornado had been very evident on the screen. Had they known what the echo represented, they might have been able to warn the city of the approaching tornado. In response to the disaster they adopted as their immediate goal the prevention of another similar calamity in the state. When Texas A&M president Marion T. Harrington saw a copy of the letter, he contacted retired navy meteorologist Captain Howard T. Orville, the first man to advocate a nationwide network of radar stations to track severe weather. The two men proposed a conference at Texas A&M to consider the possibility of creating a tornado network in Texas patterned after the Florida hurricane network.[11]

Representatives of the Weather Bureau, AWS and its SWWC, U.S. Navy, Texas Department of Public Safety, Texas Civil Defense Office, Oklahoma A&M College, University of Texas, Dow Chemical Company, Copano Research Institute, Central Power and Light Company, and the Texas A&M oceanography and electrical engineering departments met at the College Station campus on June 24, 1953, to determine whether the six existing Texas radar sites could form the core of an effective tornado warning system.[12] At the time the Weather Bureau operated radar at its offices in Wichita Falls, Amarillo, and Brownsville, and the Copano Research Foundation at Victoria, Dow Chemical at Freeport, and Central Power and Light at Corpus Christi had radar at their facilities. Four of the sites were on the Gulf Coast, however, and not in the areas most susceptible to tornadoes.

At the conference Orville traced the history of the navy's participation in hurricane warning networks. Weather Bureau participants Robert Simpson from the Washington-based Radar Program for Severe Storm Detection and Warning, Erle L. Hardy from the Southern Regional Office in Fort Worth, and Loren F. Jones from the Victoria office outlined the bureau's role in tornado forecasting and warning. Joe S. Fletcher, assistant state director of the Department of Public Safety, explained his organization's activities during natural disasters. John C. Freeman of Texas A&M's oceanography department reviewed the existing radar facilities. During the discussion session, Simpson reported that the Weather Bureau had 17 million dollars' worth of war surplus APS-2 radar sets that the air force and navy had donated. It was possible to modify the one hundred airborne radar sets to detect storms, but the bureau lacked funds for the conversions. Simpson suggested that cities with First Order Weather Bureau Offices might pay for the modifications and thus provide radar coverage for their area. The result would be a statewide radar network. Fletcher offered the cooperation of the Department of Public Safety's 450 radio-equipped cars and statewide communications network. Civil Defense director William L. McGill agreed to add his organization to the effort. Jeff Davis, editor and owner of the *Crockett Democrat*, volunteered to oversee a public education program. When all of the components were in place, Texas would have the first comprehensive tornado warning program in the country.[13]

In order for the system to work effectively, major population centers would have to participate. The Texas A&M Research Foundation, the overseer of the college's research funds, produced a brochure that described the Texas Radar Tornado Warning Network and detailed a community's cost for participation. The college's electrical engineering department would modify the radar for six thousand dollars; installation would cost an additional three to seven thousand dollars.[14] Fletcher, Hardy, Freeman, McGill, and Archie Kahan from the Texas

A&M Research Foundation addressed city councils and county commissions around the state. Fort Worth was the first to underwrite the cost of the system. After discovering their city was far down the list for receiving Weather Bureau radar, the Fort Worth City Council purchased a radar set from Texas A&M for eleven thousand dollars.[15] Shortly thereafter Houston, Galveston, Midland, San Angelo, and Abilene bought radar for their Weather Bureau Offices. By the time a second conference convened in College Station to officially dedicate the network, on June 24, 1955, sixteen Texas cities as well as Oklahoma City and Lake Charles, Louisiana, had joined the system.[16]

The radar warning system went into action whenever SELS issued a tornado forecast for any part of the state. Radar operators at the weather stations scanned their scopes for the telltale hook or comma-shaped echo. If the dreaded echo appeared, the radar operator teletyped the tornado's location, path, and speed to the Department of Public Safety Headquarters in Austin, which issued warnings to the threatened areas. Patrol cars kept the storm under observation. Meteorologists broadcasted warnings over local radio and television stations. Telephone operators and sheriff's deputies alerted rural residents, while police and volunteers notified urban dwellers of the approaching storm. The entire process, from radar observation of the tornado to implementation of the warning system, took only about one minute. A measure of the system's success was the drastic reduction in Texas tornado fatalities. With eight radar sites in operation in 1954, the system issued seventeen tornado warnings. Six tornadoes struck heavily populated areas but claimed only three victims. The following year, in spite of a record 164 tornadoes in the state, only two Texans died.[17]

Hardy, the Weather Bureau's southern regional director, told Governor Allan Shivers that completion of the network would ensure the state had "the best and most up-to-date storm warning system now available anywhere in the world."[18] But Hardy and his boss had not always been so complimentary of the project or its leaders. Reichelderfer discussed the system's "untimely" establishment in a June 30, 1953, phone conversation. The chief acknowledged that the bureau had intended to move ahead in radar storm warning, but the conference at College Station had forced it to get involved in a much larger way than originally intended. He expressed reservations about radar's effectiveness in identifying tornadoes and about dividing responsibility with a local or state organization, but Reichelderfer's greatest worry was potential loss of funding should the program fail. "It will have a bearing on the support of the Weather Bureau when appropriation time comes up next year. If we do it right and we are successful, it can be a help." In Reichelderfer's mind the situation was very similar to the problem the bureau had faced during the preceding year, when the media pressured it to begin tornado forecasts. Hardy responded that the state of Texas was going to forge ahead with the program one way or the other, but he criticized members of the college's meteorology department for being too ambitious and for starting the program without anyone's authority.[19] To diminish Texas A&M's role in the project and to assert Weather Bureau authority into the project, Reichelderfer appointed Hardy the director for Development of Radar and Storm-Warning Networks in Texas.[20] Only after the system proved its effectiveness did the bureau praise the project as "a rare and outstanding contribution of major significance to the people of the United States." It further boasted: "Not only is this the first closely knit network of radar stations set up entirely for storm detection purposes, but it is also the first example of such extensive cooperation between Federal, State, and Local agencies in the history of the Weather Bureau."[21]

The American Meteorological Society issued a position statement on radar detection of tornadoes in 1954. Its chairman, J. S. Marshall, agreed that radar could improve the effectiveness of tornado warnings but warned that "no one should assume that even the best of our storm detection radar can at present pinpoint a tornado and positively identify it as such." Marshall noted that although several radar operators had observed distinctively-shaped echoes, further study would be necessary before meteorologists could positively identify a tornado in a specific place based solely on a radar signature.[22]

Just two years later meteorologists challenged Marshall's statement. When SELS issued a tornado forecast for parts of central and eastern Texas on April 5, 1956, observers at Texas A&M turned on their radar set to watch for any approaching storms. Graduate student Stuart Bigler noticed a particularly intense echo about sixty miles west of College Station, moving rapidly eastward. When the echo developed the characteristic cyclonic hook, Bigler felt certain that a tornado was approaching Bryan and College Station. Bigler's professor, radar specialist Myron G. H. Ligda, confirmed his radar reading, and although no formal warning procedures were in place, the professor notified the police and local radio station KORA that high winds and possibly a tornado would strike the area shortly after 3 P.M. Oceanography and meteorology department chairman Dale F. Leipper warned the college authorities and the College Station School District of the approaching threat and urged them to keep students in classrooms until the danger passed. As the echo approached the cities, the hook became more distinctive, but there was no corroborating report of a tornado on the ground. The storm struck Bryan at 3:10 with eighty-five-mile-per-hour winds and one-inch hailstones. Because rain obscured their vision, observers at the college did not learn until later that two tornadoes had touched down in Bryan with no resulting casualties. The remarkable aspect of these tornadoes was not the storms themselves but that this was the first warning of a tornado based solely on radar observation.[23] Shortly after this incident, Ligda and Bigler published a report on the use of radar in severe storm detection that described characteristics of the hook echo associated with tornadoes: (1) The hook was located in the trailing half of the echo, (2) the hook in most instances developed cyclonically, (3) the echo often developed a V shape or deep indentation on the side of the echo opposite from the hooked portion, and (4) radar detected most hooked echoes within forty miles of the radar station.[24]

If radar could provide meteorologists with a valuable tool for warning the public of approaching tornadoes, the Weather Bureau would have to extend its radar network and provide its offices with better radar sets. In the last half of the decade the bureau did expand its modified World War II radar network but realized it needed a radar designed specifically for weather. The devastion caused by the tornadoes that struck metropolitan areas in 1953 and the five hurricanes that struck the east coast in 1954 and 1955 convinced Congress to approve funding increases for the Weather Bureau. Included in the appropriations bill for 1956 were funds to develop a long-range radar for meteorological purposes. Raytheon developed the designated WSR-57 in 1957. The new ten-centimeter, five-hundred-kilowatt set had a range of 250 miles and the capability of penetrating dense precipitation to return the all-important hook echo. Miami received the first WSR-57 unit in June 1959. In the next eighteen months the bureau installed the new radar at thirty offices along the Gulf and Atlantic Coasts and in the Midwest, areas susceptible to hurricanes and tornadoes. The price tag was $4.26 million.[25]

Scientists proposed various methods for spotting incipient tornadoes, but other than radar the method thought to be the most promising in the 1950s was *sferics*, a shortened term for atmospheric discharge. Over the years tornado observers frequently mentioned that lightning around a tornado appeared to have a color different from that associated with ordinary thunderclouds. This commentary led Herbert L. Jones, electrical engineering professor at Oklahoma A&M College in Stillwater, to believe that tornado-associated lightning produced a unique electromagnetic field. He began a project to determine the feasibility of detecting an incipient tornado from lightning discharge or sferic patterns in 1947.[26] Jones based his research on the notion that a cumulus cloud capable of producing a whirling funnel must have considerably more energy than an ordinary thunderstorm-producing cumulus cloud. The high-velocity updraft in the tornado cloud would increase the "rate of separation of electrical charges which in turn suggests higher electrical potentials, more energy to dissipate in each lightning discharge, and a possible increase in the number of strokes."[27] In other words, a tornado-producing thunderstorm would exhibit more lightning strokes of higher frequency than its non-tornado-producing counterpart.

To test his theory, Jones operated a superheterodyne receiver set at four hundred kilocycles per second, which was linked to a cathode-ray oscilloscope and an electrically driven sixteen-millimeter camera during the thunderstorm seasons of 1948 and 1949. Preliminary investigation of the graphs led Jones to determine that the number of strokes per unit time during the progress of a tornado increased over that for a regular thunderstorm by a ratio of about ten to one and that the amplitudes of the tornado sferics were considerably greater. When two funnel clouds passed over Stillwater in June 1950, Jones had an opportunity to test his theory further. A study of the data from these storms indicated that during the time the funnel was active the ratio of high-frequency sferics to ordinary sferics was one to one, but before and after the funnel's appearance the ratio was one to twenty. In comparison, Jones never observed a high-frequency sferic during a non-tornado-producing thunderstorm. Jones concluded that the high-energy conditions that resulted in a tornado's formation were the cause of the high-frequency sferics. The project expanded and intensified over the next six years, but the final report's only conclusion was that the number of lightning strikes per second was a function of the storm's intensity. Jones only suggested that this method could positively identify tornadoes.[28]

Although the technological advances offered promise for tornado detection, human eyes spotted most tornadoes. Thousands of communities without Weather Bureau Offices relied on local groups such as the Red Cross, Civil Defense, civic organizations, municipal governments, and farmers to organize severe weather spotter networks. Officials at bureau offices located at or near state capitals coordinated the efforts. By 1956, 172 bureau offices east of the Rockies had established local storm warning networks for their areas of responsibility. In addition to local spotter organizations, a substantial number of business and government entities, including the Civil Aeronautics Administration, Immigration Service, Army Corps of Engineers, and utility, pipeline, and oil-drilling companies, participated in warning of approaching tornadoes.[29]

One state adequately covered with spotter networks was Kansas. Wichita and Kansas City reactivated their warning systems, established in 1943, each spring. The Wichita Weather Bureau Office expanded its network to twenty-eight counties, serving a total population of more than one million people. The West Kansas Spotter-Network Warning System, organized at Great Bend in March 1956, served as an example for others throughout the country. Sergeant Alvin J. Alves of the Air Force Filter Center at Hutchinson merged elements of the Weather Bureau, the air force, law enforcement agencies, and oil companies into one observer group. The actual spotters were oil field workers in radio-equipped trucks who worked throughout the western half of the state on a twenty-four-hour basis. When they sighted a tornado or learned of a twister in their vicinity, they radioed the storm's position and track to their home base, where a dispatcher filled in a prepared form and called the air force at Hutchinson. The air force notified the Weather Bureau Office at Dodge City as well as ground observer posts within a twenty-mile radius of the funnel's reported position. Ground posts verified the tornado's progress and called police and sheriff's offices within the affected area.[30]

Whether radar or the human eye detected a tornado, individual communities had the responsibility for informing their citizens of an approaching tornado. Sociologists and psychologists have studied the problems inherent in warning of a short-lived immediate disaster such as a tornado. Waltraud Brinkmann at the University of Wisconsin wrote, "The single most important factor influencing man's response to the tornado hazard is the warning system. A warning, seemingly such a simple matter, is actually an enormously complicated process involving evaluation, dissemination, and response." A warning can only save lives if the public receives it and responds in an appropriate manner. Psychological research has determined that "the *source* of a communication is a crucial factor in determining how the receiver will respond to its substantive message."[31]

Methods of spreading the warning varied during the 1950s. In Iola, Kansas, the fire department siren, Lehigh Portland Cement Company siren, Pet Milk Company whistle, and Walton Foundry whistle would sound a five-minute continuous blast when a tornado approached.[32] Topeka, Kansas, instituted a similar system. If a tornado presented a threat to the city, whistles at the Santa Fe Railroad yards and Beatrice Food Company would emit three ten-second blasts spaced five seconds apart; they would repeat the pattern three times.[33] Cleveland, Ohio, operated the Automatic Telephone Weather Forecast WE1-1212 system. The proof of the system's success came on June 8, 1953, when a tornado moved eastward over Lake Erie toward Cleveland. Although no severe weather forecast was in effect, the Weather Bureau Office issued a warning. Seven minutes later the automated telephone system began contacting over eighty-three thousand people. Seven deaths occurred when this tornado destroyed one hundred homes on the city's west side, but loss of life might have been much greater without the warning.[34]

Radio was the most common method of disseminating tornado warnings in the 1950s. Broadcasting of weather reports over commercial radio stations began in the early 1920s. Only twenty of the nation's thirty-six commercial stations had a license to carry Weather Bureau forecasts by 1922, but the ultimate plan was for at least one station in each state to distribute official weather forecasts and warnings. The bureau certified all 140 existing stations for weather broadcasts in January 1923. Over the next few years the bureau and radio stations created links that enabled stations to cover local threats in a timely manner.[35] Most storm spotter networks phoned radio stations in their areas to report tornadoes, and the radio stations relayed the warnings to their listening audiences. Over 95 percent of American households had radios by 1950.[36]

Only 5 percent of American households had a television in 1950, although both the number of sets and stations increased dramatically during the decade. This new medium, which would become the public's chief source of severe storm warnings in the future, cautiously entered its new role in Oklahoma City in 1952, when Reichelderfer granted WKY (now KTVY), the city's sole television station, permission to use Tinker Field's tornado predictions on the air. The station also broadcast Weather Bureau tornado warnings that were confined to Oklahoma. To avoid a panic, the WKY announcer would preface all warnings or weather information with the following statement: "We will pause a moment in our regular schedule in order to bring you some important weather information." In reply to a request for listener feedback, WKY received 1,682 cards and letters. A random sample revealed the public's pleadings that they be continued.[37]

No rules existed in the early days of television weather. No ban on television or radio stations issuing their own tornado warnings existed. Those stations that rebroadcast official Weather Bureau warnings frequently did not credit the bureau as the source, causing listener confusion. Sociologists at the time found that conflicting information regarding an imminent natural disaster or advice from a source that the public deemed questionable slowed the recipient's response, often with fatal results.

When tornado forecasts and warnings became more frequent, television stations devised ways to notify the public without disrupting programming. They often displayed tornado and severe thunderstorm forecasts on station identification display cards, which they showed every half hour during commercial breaks. For warnings, which required more immediate action, their solution was to use "crawls," single lines of copy running across the bottom of the screen from right to left. Meteorologists typed messages on long strips of plastic that were mounted on a large rotating drum. Light passing through the drum illuminated the letters that appeared as a superimposed image on the bottom of the television screen.[38]

For tornado forecasts and warnings to be effective, the country's citizens had to become knowledgeable about nature's most violent storm. Various organizations participated in a public education campaign. Newspapers in Kansas and Oklahoma took the lead in featuring informative articles on tornado identification and safety precautions. The *Oklahoma City Times* published an eight-part series with accompanying photographs and drawings in 1953, and the *Topeka Capital* printed a twelve-part series in 1956.[39] During tornado season, Hoye Durham, meteorologist-in-charge at the Austin Weather Bureau Office, taped three daily broadcasts of information received from the Central Office about tornadoes and distributed the short messages to all ten radio stations in the area for transmission. The Weather Bureau Central Office printed several posters and pamphlets such as "It Looks Like a Tornado" for distribution through local offices.[40]

Taking the adage "a picture is worth a thousand words" to heart, some organizations produced short black-and-white films for public instruction. Perhaps the first instructional film on tornadoes was *Tornado Warnings*, which United World Films produced in 1952. The brief offering showed severe storm forecasting at Tinker Air Force Base and the Oklahoma City Weather Bureau Office and had footage of the 1951 Corn, Oklahoma, tornado, the first tornado ever filmed.[41] A more significant movie was *Tornado*, a fifteen-minute black-and-white motion picture that the U.S. Gas Corporation and Texas Eastern Transmission

Corporation of Shreveport, Louisiana, created in 1956. Weather Bureau Offices and state libraries loaned copies of the film to television stations, schools, and community organizations for showing. By May 1957 more than forty-two million people had seen this depiction of a tornado spotter network in action and the safety precautions citizens should take to avoid injury or even death. As a result of this educational campaign, 240 communities took action to establish a local tornado warning network.[42]

Although tornado forecasts and warnings were in their infancy during the 1950s, the system may have been the reason the tornado death rate declined substantially during the decade. One of the earliest successes came on May 11, 1953, when the New Orleans Weather Bureau Office issued a tornado forecast for parts of west and central Texas. The Highway Patrol scanned the threatening skies for a funnel and spotted one about forty miles northwest of San Angelo that was heading southeast toward the city of about fifty thousand. The patrol notified school and city officials, who passed the warning to the public. At 2:15 P.M. the tornado hit Lake View School, where students crouched in the halls in the way they had been taught during tornado drills. The storm demolished the building, but only twelve students were slightly injured. The twister destroyed or badly damaged 519 homes, leaving seventeen hundred people homeless.[43] Although the human casualties were eleven dead and sixty-six severely injured, the city's chief of police and the captain of the Highway Patrol said that in their opinion, "but for the warning a hundred more people would have been killed."[44]

Similar testimonials followed when an F5 tornado tore through the heart of Blackwell, Oklahoma, on May 25, 1955. Although twenty citizens died and 280 received serious injuries, Commissioner of Public Property E. Bagby wrote to the Weather Bureau, "Our loss of life would have been several times the above figure (20) had it not been for the broadcast warnings of the Weather Bureau. Not only were people advised of the approach of the tornado, but constant advice was given as to what precautions to take." Bank president G. E. Brambaugh similarly wrote that there was no question in his mind that without the advanced forecast and warnings, loss of life would have been considerably greater.[45]

The Kansas City metropolitan area was very fortunate that the SELS unit was located in its environs on May 20, 1957. About 6:15 that evening a half-mile-wide F5 tornado plowed a fifty-mile path of devastation through the southern Kansas City suburbs, killing thirty-seven. SELS forecaster Fred Bates later told the Kansas City branch of the American Meteorological Society that at least five hundred persons would have died in the tornado's winds, estimated then at three hundred to four hundred miles an hour, if SELS had not forecast the storm and news agencies had not cooperated in spreading the warnings. Bates based his estimate of probable deaths without warnings on interviews with persons who were able to take shelter as a result of advance notice of the storm's approach. The SELS meteorologist summarized his organization's mission: "The most precious commodity we have in this country is human life. By detecting these storms in advance and giving adequate warning, we are able to effect great savings in the national wealth."[46]

Unfortunately, the tornado forecasting and warning system in the 1950s was not consistent from place to place. On the same day that San Angelo heeded tornado warnings, Waco did not, and 114 of its citizens perished. In a similar manner, Blackwell suffered only twenty deaths, whereas eighty citizens of the neighboring community of Udall, Kansas, lost their lives the same night. Blackwell had received adequate warning; Udall had not. But the 1950s would be the last decade to report more than one thousand tornado deaths, and 1953 would be the last year more than one hundred Americans would die in a single tornado.[47]

Tornado spotting and warning programs, like tornado forecasting, were in their infancy at the beginning of the 1950s, but while one organization had the responsibility for developing tornado forecasting, numerous groups participated in the advancements made in warning the American public of tornado threats. Colleges such as Texas A&M and Oklahoma A&M were pioneers in the development of technological tools for tornado identification. Meteorologists at the Texas school, some of the first in the country to observe the characteristic hook echo pattern a tornado exhibited on a radar screen, were instrumental in establishing the first statewide radar network devoted to the detection of severe thunderstorms and tornadoes. By the end of the decade the Weather Bureau had expanded its radar capability and had begun installing a more powerful radar, the first designed specifically for meteorology. Scientists at the Oklahoma college developed a tornado detection system based on the relationship between tornadoes and lightning. To supplement these technological innovations, numerous organizations in Tornado Alley formed networks of human spotters to report tornado sightings.

Communities employed several methods, including sirens and telephone systems, to warn their citizens of a tornado's approach, but the most common method of dispersing warnings was via the broadcast media. Nearly every American household had a radio during the 1950s, but many lacked television, a relatively new invention. Although more citizens received tornado warnings from radio than from television, the new medium spread throughout the country and became a major source for storm warnings by the end of the decade. To instruct the public on actions it should take in the event of a tornado, several groups developed educational tools, such as pamphlets, spot announcements, articles, and films.

The infant tornado forecast-spotting-warning system had some successes in the 1950s as well as some devastating failures, but by the end of the decade the essential components were in place. During the next thirty-five years, advancements in forecasting techniques and technology would continue to enhance and strengthen each segment of the system in an effort to continue to reduce tornado fatalities.

CHAPTER 6

TECHNOLOGICAL IMPROVEMENTS

The decades since the 1950s have seen technological improvements in all areas of the tornado forecasting/warning system. Computers, satellites, and automated observation systems have aided severe storm forecasters since the 1960s. More advanced radar systems have helped significantly in the detection of tornadoes, even in their formative stages. Perhaps the most crucial improvements have come in the communications sector, where the expansion of television and radio into nearly every household has allowed the American public to receive notice of threatening weather and instructions on what to do in the event of an approaching tornado. In the mid-1980s, NOAA launched a $4.5 billion program to modernize the NWS based on new technology and advances in meteorology. The ten-year goal was to reduce the NWS to a streamlined network of 116 modernized high tech weather forecast offices in the U.S. The plan included four major systems: an automated surface observation system to collect meteorological data, a weather satellite system to provide cloud images from high above the earth, a new radar system to observe storms and provide timely warning of their approach, and an updated computer system to serve as the "central nervous system" in each of its offices.[1]

Processing the massive quantity of data needed to provide accurate severe weather forecasts by modern methods requires a great amount of time. To handle the enormous number of calculations, meteorologists have employed computers. Application of computers to meteorology began around the end of World War II, when mathematician John Neumann proposed building an electronic digital computer devoted to the advancement of mathematical science at the Princeton, New Jersey, Institute for Advanced Study. To demonstrate the computer's potential for solving scientific problems, Neumann chose weather prediction and established the Meteorology Project at the institute in 1946. During the next four years Neumann demonstrated that a physics-based algorithm (computer program) could predict large-scale atmospheric motions as accurately as humans could and that computers could perform calculations with enough speed and reliability to be useful in weather forecasting. Commercial computers, such as the IBM 701, became available in 1953, and by 1957 meteorologists in several countries were using computers for numerical forecasting and research.[2] By the end of the 1960s the computer had revolutionized weather forecasting. Scientists developed algorithms based on the theory that the atmosphere, a fluid, was subject to the laws of fluid dynamics, and the computers rapidly integrated the latest meteorological data into the equations to provide an estimate of the future state of the atmosphere.

The NWS implemented a large-scale computer-based program in 1978 to improve its personnel's productivity and effectiveness and to increase the timeliness and accuracy of its forecasting and warning services. The Automation of Field Operations and Services (AFOS) program provided processors and monitors to all offices according to their specific needs. Usually each forecast office received five computers, and each local office received two. The NWS distributed basic meteorological and hydrological data, analyses, forecasts, and warnings over a high-speed communications system to their offices, where both textual and graphic information would appear on the computer screens.[3]

To collect meteorological data, the NWS employed both automated and manual methods. For observations above the earth's surface, the NWS relied on radiosondes, or pilot balloons. Meteorologists at about one hundred locations around the country launched instrument packages attached to balloons at 0000 and 1200 GMT daily. The instruments relayed information on temperature, moisture, and winds at various altitudes up to one hundred thousand feet and provided information on atmospheric instability, a condition in which rising air within a thunderstorm remains warmer than the surrounding air. When air parcels are displaced from their original position, they accelerate; the atmosphere becomes more unstable.[4] Increasing instability frequently indicates increased potential for severe thunderstorms or tornadoes.

To obtain surface data the Federal Aviation Administration (FAA), private citizens, and corporations joined the NWS in staffing over one thousand weather observation stations, many located along airport runways. Where human observation was impractical, NWS used the Automatic Meteorological Observing System (AMOS) to provide the same data. As part of the modernization program the NWS began in the 1980s, a joint effort of the NWS, the FAA, and the Department of Defense (DoD) created the Automated Surface Observing System (ASOS) to replace human observers at more than 850 locations throughout the country. Atlantic City, New Jersey, the site of an FAA research center, received the first ASOS unit in August 1991. ASOS, which updates observations every minute around the clock, transmits sky conditions, visibility, barometric pressure, temperature, dew point, wind direction and speed, accumulated precipitation, and conditions that might affect visibility (fog, rain, snow) to NWS and FAA offices. The estimated cost of the program is $350 million.[5]

At NSSFC the transition from paper to electronics came in several steps. In 1975 forecasters began transmitting watches through a primitive system, essentially an electric typewriter with a CRT screen that was hooked to a communications line. Because it was not even a word processor, everything had to be spelled out in detail, but within a short time NSSFC developed templates for the watches so that forecasters had only to tab and fill in the blanks. The transmission of watches was done on computer by the end of 1976.[6]

Interactive computer capability began at NSSFC in December 1980, when the center piggy-backed onto the Man-Computer Interactive Data Access System at the University of Wisconsin. Forecasters were able to access satellite information and perform data analyses in real time. NSSFC installed its own standalone computer system, the Centralized Storm Information System (CSIS), in February 1982. CSIS, which consisted of a geostationary operational environmental satellite (GOES) receiving antenna, three Harris/6 computers, three interactive terminals, and access to FAA observational data and radar data, allowed much quicker retrieval and analysis of real-time data. In 1989 the NSSFC implemented a more powerful computer system, the VAS Data Utilization Center (VDUC), which allowed the inclusion of additional data sets such as lightning detection and wind profilers.[7]

To replace the aging AFOS system, in the mid-1980s the National Oceanic and Atmospheric Administration (NOAA) proposed a high-speed computer workstation and communication network that could rapidly integrate the data from ASOS, GOES, and Doppler radar. Still to be completed, the new Advanced Weather Interactive Processing System (AWIPS) will cost over $525 million and will "support forecasters in graphically integrating and analyzing the volumes of weather observations and products that form the basis for decisions on each day's forecasts and warnings." Installation began in September 1997. When completed in 2003, AWIPS will be the nerve center of operations at the Storm Prediction Center at Norman, Oklahoma, and all 119 modernized Weather Forecast Offices.[8]

Another technological advancement that aids severe weather forecasting is satellite photos. The use of cloud observations to forecast weather is a long-established practice. Over the centuries, weather proverbs such as "mackerel clouds in sky, expect more wet than dry" and "a curdly sky will not leave the earth long dry" noted the clouds' importance in predicting future weather.[9] Satellites made cloud observations over large expanses of territory practical. The meteorological satellite era began on April 1, 1960, with the launch of TIROS-1, which used shuttered television cameras to record cloud images on tape for later transmission to the ground. During the next five years the U.S. armed forces, the National Aeronautics and Space Administration (NASA), and the Department of Commerce launched ten experimental TIROS satellites. Nine satellites, named ESSA-1 through ESSA-9 for the Weather Bureau's parent agency, the Environmental Science Services Administration,[10] provided routine global coverage. A breakthrough in severe weather forecasting came with the NASA launch of the first Applications Technology Satellite (ATS-1) on December 6, 1966. The ATS-1's spin-scan cloud camera provided visual images of the earth and its cloud cover every twenty minutes. Now meteorologists had views of clouds and cloud systems in motion. They realized that they could use these images to infer wind velocities at the clouds' altitudes. NSSFC began using satellite movie loops in the spring of 1972.[11]

NASA launched a satellite that remains in orbit above the same spot on earth (geostationary) in 1974, and the following year NASA placed NOAA's first GOES in orbit. During the next twenty years eight more satellites were launched to replace malfunctioning satellites or upgraded antiquated equipment. Geostationary satellites orbit in the earth's equatorial plane 35,800 kilometers above the earth's surface. Because at this altitude their west-to-east motion equals that of the earth, they remain stationary at the desired longitude. The GOES West, located at 135 degrees west longitude, and the GOES East, at 75 degrees west longitude, provide satellite coverage of the central and eastern Pacific Ocean, North, Central, and South America, and the central and western Atlantic Ocean. The annual estimated cost of the GOES satellite system is $2 billion.[12]

Geostationary satellites allow meteorologists to watch developing cloud patterns even before precipitation begins and often give the only indication of developing storms, but satellites are not as useful for forecasting fleeting mesoscale phenomena as they are for forecasting large-scale weather features such as fronts and hurricanes. To increase the satellite's usefulness to severe weather forecasters and to explore the feasibility of detecting tornadic signatures on satellite photos, Charles Anderson, professor emeritus of marine, earth, and atmospheric science at North Carolina State University, developed the Detection of Tornadic Thunderstorm (DOTT) project in 1992. DOTT researchers compared two consecutive close-up satellite pictures of individual thunderstorm cells to detect and evaluate changes in anvil cloud tops.[13] Strong updrafts present in tornadic cells often push the tops of anvil clouds downstream from the cloud itself or cause anticyclonic rotation in clouds near the updraft column. Based on these aberrations the DOTT computer program, which assigned each thunderstorm a tornado probability and strength value, identified about 90 percent of all tornadoes that year. Lead times were unfortunately very short, especially for strong tornadoes, which often develop very quickly. These storms frequently formed during the thirty-minute gaps between satellite photos. In spite of its limitations, Anderson believed his program could help tornado forecasting in areas where radars were out of operation or coverage was minimal, but expansion of Doppler radar coverage negated the program's usefulness.[14]

Although the NWS has made tremendous advances since the 1950s in collecting and processing data to aid the weather forecaster, actual day-to-day tornado forecasting techniques have changed little. Tornado forecasting still remains in human hands and depends on the severe weather forecaster's knowledge and experience. Stephen Corfidi, a lead forecaster at the SPC, likened his job to medicine: the meteorologist uses data, observations, and experience to diagnose a weather situation much as a doctor uses the same criteria to diagnose a patient. Occasionally the diagnosis is wrong, not because either one misinterpreted the data but because each storm and each person is unique and does not always respond in the same manner as hundreds of previous storms and patients. Though checklists of parameters exist in the back of his mind, the SPC forecaster adds experience and intuition to the scientific data to decide if and when he should issue a tornado watch.[15]

The process of formulating a tornado watch at the SPC may begin up to two days before the actual issuance. After studying past and current data, recent weather trends, and computer forecast models, outlook forecasters prepare a Day-1 and a Day-2 Outlook. The Day-1 Outlook, issued five times daily, outlines areas within the United States where severe thunderstorms may develop during the next six to thirty hours. Similarly, the Day-2 Outlook, issued twice daily, covers the succeeding twenty-four-hour period. Both outlooks plot a risk area and evaluate the possibility of thunderstorm development, ranging from nonsevere to high risk for severe thunderstorms. For areas designated as high risk, which indicates a potential for a major severe weather outbreak, including violent tornadoes, the SPC issues a public information statement describing the particularly dangerous situation. Although the SPC does not announce these outlooks to the public, they are available on the Internet and are a good notification to spotters that they would be wise to keep in touch with their local network coordinator.

When conditions for severe weather begin to develop, the mesoscale forecaster, who specializes in forecasting severe weather in a time frame of up to six hours and in an area the size of several small states, issues a mesoscale discussion statement (MCD) anywhere from half an hour to several hours before issuing a watch. The MCD addresses areas of current or expected severe activity and indicates whether or not severe thunderstorm or tornado watches might be forthcoming. The MCD gives local NWS offices advance notice to begin preparing for severe weather. The lead forecaster, who supervises all activities on his or her shift, has the ultimate responsibility for issuing tornado watches. The decision to issue a tornado watch is made after the lead forecaster consults the mesoscale forecaster and the local NWS offices and is based on meteorological data and past experience.[16]

A watch alerts the public and local NWS offices to increasing danger of severe weather, activates local storm-spotter networks, and alerts meteorologists in the watch area to closely monitor radar screens for signs of developing tornadoes. For forty years tornado watches have followed a basic format that includes the issuing agency, the time and date of issuance, the areas of the country for which the watch is valid, the effective times for the watch, a reminder to the public of the watch's meaning, and the forecaster's name. The few changes that have occurred include the shift from a typewritten form to a computer program and the institution in 1966 of the term *watch* for the previously used term *forecast*. The following is a sample watch:

BULLETIN—IMMEDIATE BROADCAST REQUESTED
TORNADO WATCH NUMBER 547
STORM PREDICTION CENTER NORMAN OK
1229 PM CDT TUE JUL 1 1997

THE STORM PREDICTION CENTER HAS ISSUED A TORNADO WATCH FOR PORTIONS OF EASTERN SOUTH DAKOTA

EFFECTIVE THIS TUESDAY AFTERNOON AND EVENING FROM 100 P.M. UNTIL 700 PM CDT

TORNADOES . . . HAIL TO 2 INCHES IN DIAMETER . . . THUNDERSTORM WIND GUSTS TO 75 MPH . . . AND DANGEROUS LIGHTNING ARE POSSIBLE IN THESE AREAS.

> THE TORNADO WATCH AREA IS ALONG AND 70 STATUTE MILES EAST AND WEST OF A LINE FROM 40 MILES NORTH NORTHEAST OF ABERDEEN SOUTH DAKOTA TO 25 MILES SOUTHEAST OF MITCHELL SOUTH DAKOTA. REMEMBER . . . A TORNADO WATCH MEANS THAT CONDITIONS ARE FAVORABLE FOR TORNADOES AND SEVERE THUNDERSTORMS IN AND CLOSE TO THE WATCH AREA. PERSONS IN THESE AREAS SHOULD BE ON THE LOOKOUT FOR THREATENING WEATHER CONDITIONS AND LISTEN FOR LATER STATEMENTS AND POSSIBLE WARNINGS.
>
> WEISS

When major outbreaks are a possibility, the watch may include the additional wording, "This is a particularly dangerous situation with the possibility of very damaging tornadoes. Also . . . large hail . . . dangerous lightning and damaging thunderstorm winds can be expected." SPC is developing software that will allow the issuance of tornado watches by county or parts of a county. This will provide more flexibility than the rectangular watch boxes.

Research is an essential companion to severe weather forecasting success. A research program for severe storms, separate from SELS, had its inception when a panel discussion on tornadoes at the November 1954 American Meteorological Society meeting suggested the use of aircraft to observe and measure certain mesoscale features associated with severe storms. An airplane would be less costly than an extensive upper-air sounding network and would provide a better picture of a thunderstorm's three-dimensional structure. At the next year's AMS meeting Clayton F. Van Thullenar, director of the Weather Bureau's Kansas City regional office, SELS supervisor Donald House, and SELS researcher Robert Beebe met with James M. Cook, an experienced pilot and cloud seeder. The meteorologists wanted Cook to fly his specially equipped F-51 airplane into tornado forecast areas to take temperature and humidity readings. Cook signed a contract with the Weather Bureau in November 1955 to make observations of temperature and humidity gradients in both the horizontal and vertical plane before, during, and after thunderstorms and tornadoes. The research program, named the Tornado Research Airplane Project (TRAP), conducted flights in and around thirty-six tornado and severe thunderstorm forecast areas on thirty-three days during the spring and summer of 1956. SELS hired several extra people to process the volume of data the flights collected.[17]

To provide a more concentrated effort in this area of research, in 1959 the Weather Bureau established within SELS a new unit, the National Severe Local Storms Research Project (NSLSRP; renamed the National Severe Storms Project—NSSP—in 1960), and named Van Thullenar director. The bureau collaborated with the air force, the navy, the FAA, and NASA to fly multiple-plane missions into thunderstorms and squall lines during the spring of 1959 and 1960. Although the project's headquarters were in Kansas City, the base of operations was Oklahoma City. The bureau's WSR-57 radar at Oklahoma City's Will Rogers Field was critical for both safety and research analysis.

Conflicts arose over the use of the radar when the Weather Bureau had to give priority to its public forecasting and warning programs over pure research. To solve the problem, the bureau installed another WSR-57 on the North Campus of the University of Oklahoma at Norman in 1962. This site, designated the Weather Radar Laboratory, became the logistical and communications base for NSSP's fieldwork. Robert H. Simpson, the bureau's director of severe storm research in Washington D.C., realized that the entire program would operate more efficiently from Norman, where students and faculty at the University of Oklahoma's meteorology and electrical engineering departments could provide additional input and personnel. In addition, Norman's geography was better suited to field studies than were areas around Kansas City. The NSSP relocated in Norman to 1964 under the direction of Edwin Kessler, formerly a meteorologist with the Travelers Research Center in

Hartford, Connecticut. To emphasize its devotion to research, and the fact that it does not forecast severe weather, the NSSP was renamed the National Severe Storms Laboratory (NSSL). NSSL's objectives were to increase knowledge of severe local storms in order to contribute to improved forecasting; to develop new and more effective means for collecting and analyzing severe storm data; to develop equipment, especially radar; and to improve operating methods of both personnel and equipment in order to provide more timely and accurate information to severe weather information users.[18]

From its inception the laboratory gave high priority to the development of an effective Doppler radar program. Although conventional weather radar could determine a storm's location, movement, and relative intensity, it did not have the capability to determine the variation of motion of water, ice, or dust particles in the atmosphere, a good indicator of the type of weather that was developing. On the other hand, Doppler radar could measure both wind velocity and direction around atmospheric particles. The Doppler system is based on the Doppler effect, a frequency shift Austrian physicist Christian Doppler (1803–1853) recognized in 1842. Doppler observed that a train whistle's pitch was higher as the train approached a listener and lower as it receded. An approaching sound source compressed the sound waves, thus producing a higher frequency, whereas a receding sound source stretched the sound waves, lowering their frequency. A radar based on this principle could measure the velocity of anything, even dust particles, moving toward or away from it and could give meteorologists knowledge of wind movement even when clouds or rain were not present.[19]

One scientist to recognize the potential for Doppler's use in meteorology was James Q. Brantley at Cornell Aeronautical Laboratory, who in 1957 presented the first report of Doppler measurements of weather echoes at the Sixth Weather Radar Conference of the American Meteorological Society. The paper included an analysis of a continuous wave Doppler radar spectra of rain, snow, and thunderstorms under different conditions. Though this work marked the beginning of a new area of radar meteorological research, the lack of a pulse radar Doppler hindered research. Continuous wave radar cannot determine range as a pulsed radar does. Because the transmitter and the receiver are located in the same place, the radio waves from a continuous wave radar can interfere with the returning waves. Pulse radar, which emits the radio waves in intermittent bursts perhaps only a millionth of a second in duration, waits a few thousandths of a second to receive the returning waves. With range information provided and interference reduced, the pulse radar gives a much more complete picture.[20]

The Weather Bureau obtained a continuous wave Doppler radar from the U.S. Navy in the fall of 1956 and mounted it on a mobile van. Throughout the tornado seasons of 1957, 1958, and 1959 severe storm researchers used it to locate and identify tornadoes in Kansas and Texas. Their first success occurred at El Dorado, Kansas, on June 10, 1958, when they identified a tornado based on its appearance on a Doppler radarscope. The radar measured the tornado's wind speed at 205 miles per hour.[21]

Roger M. Lhermitte, a radar specialist who had helped develop the three-centimeter Doppler the French built in 1956, joined NSSL in 1964. He and other NSSL engineers under his direction built a similar three-centimeter Doppler radar, which they used for research until they converted surplus air force early-warning radars into two ten-centimeter pulsed Dopplers specially designed for meteorological purposes. The first NSSL ten-centimeter pulsed Doppler radar became operational at Norman in 1971. With the new radar meteorologists had little difficulty identifying a mesocyclone, the parent of many tornadoes. Graham Armstrong, a radar engineer at the Air Force Geophysics Laboratory in Cambridge, Massachusetts, devised a Doppler radar that could indicate wind shear patterns in 1966, and for two years he collected pictures of the radarscopes during thunderstorms in the area of the laboratory. After a study of the data, radar meteorologist R. J. Donaldson at the Air Force Cambridge Research Center identified the existence of a distinctive Doppler mesocyclone signature or radar pattern and proposed minimum values for amount of shear, height of shear, and persistence of the mesocyclone. At NSSL, Don Burgess conducted studies on mesocyclone signatures and, using Donaldson's criteria, identified thirty-seven mesocyclones on the laboratory's Doppler radar. Twenty-three of the mesocyclones were associated with tornadoes and, more important, no verified tornado occurred without a preceding mesocyclone signature. Burgess found an average lead time between mesocyclone detection and tornado touchdown of thirty-six minutes.[22]

Another major breakthrough in tornado identification by Doppler radar occurred on May 24, 1973, when a tornado struck Union City, Oklahoma, only a few miles from NSSL. From the mass of data and radar pictures this storm provided, the Norman researchers discovered the radar signature of the tornado vortex itself. A tornado vortex signature manifests itself on Doppler radar as "extreme localized wind shear over a distance of about one kilometer."[23] In other words, a tornado exhibits different wind velocities and directions within a very short radius, and Doppler radar is able to "see" these differences.

To test Doppler radar's capabilities in an operational environment, the NWS, NSSL, FAA, and the Air Force Geophysics Laboratory conducted the Joint Doppler Operational Project from 1976 to 1978. The Doppler system accurately identified mesocyclones at distances up to 250 kilometers and tornadoes up to 125 kilometers from the radar site. Convinced of Doppler's effectiveness in giving advance warnings of severe storms and floods, the NWS, FAA, and AWS agreed that they needed the new type of radar to replace the aging WSR-57 sets. The president's budget for 1981 included funds for an extensive network of Doppler radars, named the Next Generation Radar or NEXRAD, which took several years to perfect. NWS installed its first operational NEXRAD unit, officially known as the WSR-88D, at the Oklahoma City Weather Service Office in May 1990 (in 1970 the Weather Bureau became the National Weather Service). During the next six years the NWS, FAA, and DoD installed an additional 163 NEXRAD units at sites throughout the United States, Puerto Rico, and Guam.[24] Doppler's greatest benefit is the increased lead time it provides in tornado warnings. The average lead time a Doppler-based tornado warning provided in 1996 was 17.0 minutes. By comparison, the average lead time a storm-spotter tornado warning provided for the same year was 12.0 minutes.[25] Of course, Doppler radar detection of mesocyclonic circulation alerted many of these spotters to the possibility of a tornado.

Another type of Doppler radar is the wind profiler, which points vertically to provide wind speed and direction readings for both horizontal and vertical winds directly above the radar site. Between 1986 and 1992 the NWS installed thirty-two wind profilers across the country at a cost of $53 million. The profiler sites lie between radiosonde sites, thus supplementing the weather balloon's upper-air measurements. The radar senses wind changes between five hundred meters and sixteen kilometers above the surface. The greatest benefit for severe storm forecasters is the system's ability to detect local wind shear patterns that might signal the development of a severe thunderstorm or tornado.[26]

By the mid-1990s NEXRAD had become the most widely used method of tornado detection. Paul Polger, whose office at NWS headquarters collected information on the basis for tornado warnings, reported that in 1996, 87 percent of the tornado warnings were based on Doppler radar.[27] But Doppler radar can detect only the potential for severe weather. A tornado vortex signature on the radar screen indicates the presence of wind circulation patterns associated with many tornadoes, but the radar cannot show whether a vortex is on the ground or aloft. In addition, a tornado may develop in areas without adequate Doppler coverage. The only reliable method of verifying a tornado's existence is eyewitness observation.

During World War II the Weather Bureau established a ground observer network to scan the skies around military establishments for tornadoes. The program continued in some parts of the country, especially the Great Plains, throughout the next two decades. These spotters or observers, however, had little training beyond movie clips of tornadoes and instructions for contacting their local Weather Bureau Offices. After the devastating April 11, 1965, outbreak, which took the lives of 256 Americans in six Midwest states (Illinois, Indiana, Iowa, Michigan, Ohio, Wisconsin), a Weather Bureau survey team recommended the establishment of a National Disaster Warning Program. An important component of this system, a coordinated spotter network consisting of thousands of individuals and organizations, began operation throughout Tornado Alley in January 1969. Designated SKYWARN, the spotter network provides timely reports of severe weather, especially tornadoes, from radio- equipped vehicles and homes. Unlike their earlier counterparts, SKYWARN members receive training through NWS-sanctioned programs. Personnel at local NWS offices began presenting slide and movie lectures in the 1970s, and the NWS printed spotter's field guides and other brochures for mass distribution. Each year thousands of volunteers attended training sessions to receive SKYWARN certification. Local SKYWARN members begin operation whenever the SPC issues a severe thunderstorm or tornado watch for an area or whenever the local NWS office believes threatening conditions exist within its jurisdiction.[28]

Although the spotter network and Doppler radar provide adequate warning in most cases, scientists have continued to search for additional tornado detection methods. In 1969 Newton Weller, a private researcher in Iowa, described a procedure that operates on the principle that tornado-producing thunderstorms produce electromagnetic radiation of a certain frequency. Weller said that if a person tuned his or her television set to channel 13, darkened the screen, and then quickly turned to channel 2, a glowing white screen would indicate the presence of a tornado within forty kilometers. Unfortunately, the method would not be successful every time because some severe thunderstorms do not produce enough radiation for channel 2 to detect.[29]

Some scientists continue to explore offshoots of Herbert Jones's sferics theories of the 1950s. They believe a strong correlation exists between the frequency and intensity of cloud-to-ground lightning strikes and tornado formation. To aid in this research, as well as to provide pertinent information to agricultural and business clients, the NWS has become a major customer of Global Atmospherics Incorporated's National Lightning Detection Network (NLDN). The system of over one hundred ground-based sensing stations monitors cloud-to-ground lightning activity across the country and provides real-time data to subscribers. Researchers use NLDN's database to study patterns that might lead to a method of determining tornado formation, but so far they have uncovered no conclusive relationship between the two atmospheric phenomena.

In a similar vein, NASA launched into orbit in April 1995 the optical transient detector (OTD), an instrument designed to detect and record lightning within a cloud or between clouds. This type of lightning is not visible from the ground, and the NLDN does not detect it. Dr. Hugh Christian, the principal investigator of the OTD at Marshall's Global Hydrology and Climate Center in Huntsville, Alabama, noted that when the OTD passed over a tornado-producing storm in Oklahoma on April 17, the lightning flashes peaked and then dramatically decreased just before the tornado touched the ground. The OTD detected almost two hundred flashes during a three-minute pass over the area. In contrast, the NLDN located only nine flashes during the same period. This finding led meteorologists to suggest that a high rate of intracloud lightning might suggest tornado formation, which, if true, could lead to development of new tornado detection methods, but results have proved inconclusive.[30]

An entirely different method of tornado detection focused on tornado-generated seismic activity. Although Doppler radar can identify the tornado circulation at cloud level, the only accepted method for determining that a tornado is on the ground is human observation. Researchers at the University of Alabama asserted in 1994 that a tornado in contact with the ground produced a significant seismic signal that a seismograph could detect. They based their research on the fact that a tornado encounters considerable frictional resistance with the ground, which in turn causes the tornado to transfer a considerable amount of energy to the ground in the form of seismic vibrations. The researchers proposed employing the seismologic monitoring stations that already exist for earthquake detection and oil and gas exploration to detect a tornado seismic signal. The system would be especially valuable in the southeastern United States, where low clouds and trees frequently obscure a spotter's view. Research on this method is still in progress.[31]

Al Bedard, a researcher at NOAA's Environmental Technology Laboratory at Boulder, Colorado, was testing an acoustic system designed to detect avalanche sound waves in 1995 when he picked up signals from a tornado fourteen miles from the research site. He described "the first, clear experimental evidence that sound waves were generated by atmospheric vortices" as a complete surprise. Tornadoes generate sounds in the range of one to five hertz, much too low for humans to hear, but the avalanche observing system has detected the sounds of twenty-seven tornadoes, some as far away as one hundred miles. Bedard summarized the goal of tornado researchers as the creation of a small, inexpensive tornado detector, suitable for use in the home, that would provide residents with time to seek cover or evacuate.[32]

Although forecasting and detection are indispensable components of the entire system designed to save lives, the public rarely has contact with or even gives thought to the meteorologists and scientists involved in these activities. For the public, the most important link in the tornado forecast/warning system is the entity that issues the warning. In the early days of tornado forecasting radio was the only source for warnings for most Americans, but television's rapid expansion in the 1950s allowed the new medium to join radio as the purveyor of severe weather information. According to the 1960 U.S. Census, 91.5 percent of American households had radios and 87.3 percent had at least one television set. In some tornado-prone areas television ownership was even higher: For example, 89.9 percent of the homes in Topeka, Kansas, and 92.3 percent of the houses in Wichita Falls, Texas, had a television set in 1960.[33]

The NWS is legally responsible for forecasting and detecting developing storms, for formulating warning messages, and for distributing these warnings to broadcast facilities, but no law or regulation requires any radio or television station to broadcast the tornado warning messages it receives from the NWS. Participation in the warning process is voluntary and varies greatly from one locale to another.[34] All tornado warnings must contain the name of the local NWS office issuing the warning, the effective time for the warning, the county or counties for whom the warning is in effect, and precautions for citizens to take. Each office is free to include additional information, such as communities in the storm's path, expected time of the storm's arrival at various locations, and the source of the tornado's identification. For example, the Houston/Galveston, Texas, office issued the following warning:

> Bulletin—EAS Activation Requested
> TORNADO WARNING
> National Weather Service Houston/Galveston TX
> 328 PM CDT THU OCT 23 1997

> The National Weather Service in Houston/Galveston has issued a tornado warning effective until 430 PM CDT for people in the following location . . .
> IN SOUTH CENTRAL TEXAS . . . AUSTIN COUNTY
> At 326 PM . . . Radar indicated a tornadic thunderstorm 10 miles southwest of Bellville. This storm was moving to the northeast at 30 mph. Bellville and Cat Spring are in the path of this storm. Do not use your vehicle to try to outrun a tornado. Vehicles are easily tossed around by tornado winds. If you are caught in the path of a tornado . . . leave the vehicle and go to a strong building. If no structure is nearby . . . seek shelter in a ditch or low spot.

Unfortunately, not all local NWS offices issue to the public the precaution instructions, known as the "call to action" statement, with their warnings. These statements include admonitions such as "abandon cars and mobile homes for a sturdier building or get into a ditch or culvert" and "seek refuge in a small interior room such as a closet on the lowest floor," as well as other precautions those in the path of a tornado should take. The NWS office that issues the tornado warning may or may not attach these safety statements, which are part of the Warning and Interactive Statement Editor software program the offices use for preparing warning statements. The meteorologist need only run down the list of available options and choose the ones most suitable for his or her particular locale.[35] A random survey of these statements that the NWS offices across the country issued in 1997 and 1998 showed that the calls to action varied greatly from one office to another. Occasionally, they omitted the statements altogether. More frequently, they issued inappropriate calls to action, such as giving instructions on what to do in case of a severe thunderstorm (this was attached to a tornado warning) or telling those in rural areas to go to the lower interior floors of high-rise buildings. Most critical, though, was the failure of offices such as those in Memphis, Tennessee, and

Jackson, Mississippi, whose areas of responsibility have numerous people living in mobile homes, to warn these residents to seek shelter in more substantial structures. Forty-five percent of 1997 tornado fatalities and 50 percent of 1998 tornado deaths occurred in mobile homes. NWS offices in the South, which has more than half of the nation's occupied mobile homes, issued specific calls to leave these structures only 29 percent of the time during March, May, and October 1997 and only 25 percent of the time during May and June 1998. On the other hand, the Norman, Oklahoma, NWS office, which issues more than one thousand tornado and severe thunderstorm warnings annually, during the same period always included the call to action to leave mobile homes and vehicles.[36]

Another determining factor in whether a particular location receives an adequate and informative warning is the local television station. Volunteerism and independence from the NWS led some television stations to treat the potentially life-threatening situations in a haphazard or flippant manner, while others, especially those in the tornado-prone Midwest and Great Plains states, excelled in their severe weather response. When a hook echo appeared on the Weather Bureau's radar screen in Wichita Falls, Texas, on April 3, 1964, station KAUZ's cameras began a live scan of the skies. As a spectacular twister roared through parts of the city, the station rewarded its viewers with the first live television broadcast of a tornado.[37] After a series of tornadoes raked Minnesota in 1964, Minneapolis's KSTP built an Emergency Weather Center, a separate studio complete with weather charts, lights, and audio and video equipment. With a push of a button the meteorologist could cut into regular programming to warn the public of approaching storms.[38] Perhaps station WIBW in Topeka, Kansas, best exemplified the increasingly important role television played in warning citizens of impending disaster. The station received a report over the Kansas Statewide Weather Wire at 6:37 P.M. on June 8, 1966, that a tornado was about seventy-five miles west of the Kansas capital city. WIBW staffers rushed to designated observation points, and within twenty minutes photographer Ed Rutherford reported to the station by two-way radio that a tornado was approaching the city from the southwest. Several minutes before the Civil Defense sirens sounded, newscaster Bill Kurtis began warning Topeka residents to take the necessary precautions. As the tornado ripped a twenty-mile-long path diagonally across the heart of the city, Kurtis excitedly pleaded, "If you're not undercover now, for God's sake, take cover!" Disaster researchers gave much of the credit for keeping the death toll to seventeen and for averting panic to WIBW.[39]

By the 1970s many large and midsize stations in the eastern two-thirds of the country were operating their own radar units. Those unable or unwilling to purchase their own radar could connect to the local NWS radar site. To aid in television display and attract viewers, some large-market stations introduced color radar, which gave a better indication of the precipitation's intensity than did the traditional black-and-white displays.[40] The NWS established a set of five colors, ranging from light green to red, to correspond to the radar's reflectivity, or amount of energy being returned to the radar site. The deeper and brighter colors correspond to greater reflectivity, which means the particles in the clouds (rain, snow, ice, hail) are larger and more concentrated. A red or yellow area on a radar screen indicates intense rain or hail and is often an indicator of a severe thunderstorm or a tornado.

Doppler radar began replacing traditional radar at television stations during the 1980s. Through the prodding of chief meteorologist Gary England, Oklahoma City's KWTV installed a Doppler radar system in May 1981. The same year KSTP in Minneapolis became the second station to have Doppler. When severe weather threatened south central Oklahoma on March 15, 1982, England broke into programming to broadcast the first television Doppler radar-based tornado warning. Pointing to the Doppler display of wind speed and direction, England warned the residents of Ada that a tornado would be in their area within the hour. The tornado destroyed a mobile home park, leaving one fatality behind, but the number of lives the warning saved was unknown though no doubt substantial.[41] In spite of the great potential to save lives by giving adequate lead time for residents to take shelter, by 1985 only twenty-six stations, most in very large markets or in areas highly susceptible to tornadoes, had installed Doppler radar. Two factors—the $1 million price tag and the complicated radar display—hindered rapid acceptance.[42] Vast improvements to the displays, training of meteorologists in the use and interpretation of Doppler radar, and the public's growing familiarity with the new radar led most television stations to acquire their own Doppler radar sets by 1995.

England helped develop two computerized systems to warn viewers of severe weather watches and warnings and to project a storm's path. First Warning software, introduced in January 1991, identifies the specific counties for which either a tornado watch or warning is in effect and colors the counties on a U.S. Geological Survey map, which the station displays in a corner of the television screen. First Warning also automatically generates a screen crawl. A majority of the television stations in the Great Plains, Midwest, and South have installed First Warning or a similar system. Stormtracker, which KWTV first used in March 1990, predicts a tornado's path. The meteorologist spots the storm on radar and notes its direction of movement and forward speed. The Stormtracker program plots the information on a map and indicates the localities in the path and the estimated time of the storm's arrival at each point.[43]

Tornadoes do not always strike during the hours local television stations are on the air, and if stations stay on all night, meteorologists are not always in the studio. To take advantage of this situation and the public's increasing desire for severe weather information, John Coleman created the Weather Channel in 1982. This round-the-clock cable channel, headquartered in

Atlanta, began with 4.2 million subscribers, but by its tenth anniversary Nielsen ratings certified that the network had surpassed fifty million subscribers and reached 54.7 percent of all households with television. One of the Weather Channel's greatest services concerning severe weather is the special warning scrawl on a red background. Local television station computers pick up any warning a local NWS office issues and display it in the affected locale for as long as the warning is valid to draw immediate attention. Across the nation, travelers faced with potential severe weather but unfamiliar with county names or maps that appear on local television stations have often relied on the Weather Channel for their severe weather warnings.[44]

Although most Americans rely on commercial television or radio for their tornado warnings, the NOAA began an effort in 1975 to broadcast weather information and severe weather warnings from each local NWS office. The Weather Bureau had initiated continuous weather radio broadcasting in Chicago in 1953, and during the next twenty years the direct broadcast system expanded along the coasts to provide current weather and storm information to the marine community. Additional funding in the early 1970s allowed the NOAA to begin a nationwide expansion of the FM weather radio network. The White House designated NOAA Weather Radio as the sole government operating system to provide direct warnings of a natural or nuclear disaster to private homes in January 1975. By 1997 more than 450 transmitters in the fifty states, Puerto Rico, the U.S. Virgin Islands, and the U.S. Pacific Territories, each with an effective range of forty miles, were airing taped NWS forecasts and local weather conditions on seven VHF-FM frequencies between 162.400 and 162.550 megahertz. These frequencies, outside the range of public AM and FM bands, require special radios for reception.[45]

NWS offices tailor their broadcasts to local needs. During severe weather, forecasters interrupt routine messages with tornado warnings. The forecaster transmits a special signal that activates alerting devices on many weather radios. The audible tone or visual alarm indicates that an emergency condition exists within that particular transmitting station's area and alerts the listener to stay tuned for information. After a tornado killed twenty members of an Alabama church during Palm Sunday services in 1994, Vice President Al Gore began a campaign to ensure that every public gathering place, from schools, churches, and hospitals to shopping malls and factories, had a weather radio. Additionally, NOAA agreed to install enough transmitters to ensure that at least 95 percent of the American public had access to NOAA Weather Radio.[46]

Many communities warn their citizens of approaching tornadoes by sounding a fire or Civil Defense siren. Usually the community uses a unique signal to distinguish a tornado warning from a fire or nuclear attack warning. Carrollton, Texas, has seventeen large, rotating warning sirens located throughout the city. A three-minute steady tone indicates that a trained observer or radar has determined a tornado is in the area and residents should take cover.[47] Unfortunately, in many towns the sirens, which were installed in the 1950s as part of the Civil Defense plan to warn of nuclear attack, either do not work properly or are not adequate to warn the entire population, and often a loss of power negates their effectiveness. Nevertheless, many people in tornado areas, especially those in rural areas with limited television accessibility, rely on sirens for their warnings.

The most elaborate tornado watch/warning system in the world cannot prevent loss of life unless the public knows how to react to severe weather threats. Several organizations, including the NWS, state and local disaster preparedness offices, the Red Cross, the schools, and the media, have shared the responsibility since the early 1960s for educating American citizens on actions they should take when a tornado threatens their community. All NWS-issued tornado watches and warnings include instructions on precautions to take. Many local Weather Service Forecast Offices in the 1970s designated one meteorologist as the disaster preparedness meteorologist, whose job was to work with civil defense groups, engineers, and scientists to prepare weather safety information; distribute severe storm publications, slides, films, and news releases to inform the public; and work with state officials to encourage tornado drills in schools.[48] Since the late 1950s the NWS has produced tornado safety brochures, which it periodically updates, for public distribution and tornado rules posters for display in all public facilities.

Although tornadoes have struck schools during school hours very infrequently, the possibility of such an occurrence motivated states and communities to require tornado drills and to incorporate lessons on tornado safety into their science curriculum.[49] To aid in this endeavor, the NWS produced coloring and activity booklets such as *Tornado Warning: Owlie Skywarn* and *Billy and Maria Learn About Tornado Safety*,[50] and the Weather Channel created "Weather in the Classroom" segments on tornadoes specifically for classroom use. NWS and local television meteorologists, as well as groups interested in severe storm spotting and education such as the Texas Severe Storms Association, frequently visit schools to teach the students about tornadoes safety.

The television and news media may be the best tornado education providers simply because more people watch television or read newspapers than pick up brochures. During their state's severe weather week, local stations often broadcast tornado videos of their own creation or more polished and scientifically oriented offerings such as *Nova's* "Tornado" or *National Geographic's* "Cyclone." During tornado season, newspapers in tornado-prone areas take a major part in severe weather education. Doris Ann Ware of the *Omaha World-Herald* described the advantage the print media had over the broadcast media: A newspaper could give thorough, detailed information that readers could clip and save for future reference. The *Herald* urged readers to cut out the lists of tornado safety tips and post them on prominent places throughout the house, to discuss the list with every family member, and to hold a severe weather drill.

Whether or not the public receives a timely warning and knows what to do to prevent loss of life may not be the only factors involved. Sociologists, psychologists, and others who have studied individuals' responses to warnings have identified several reasons people fail to heed the warnings. Psychiatrist John H. Sims and geographer Duane D. Baumann published an article in *Science* in 1972 that suggested that tornado death rates in Illinois and Alabama might be related to the psychological makeup of the two states' residents. Their study concluded that Illinois residents were more prone to take responsibility for their own lives, to "use their heads" and the available technology to make decisions, and to take positive action to seek shelter in the face of a tornado threat. On the other hand, Alabama residents were more likely to believe that God controlled their lives, less likely to put trust in science, and more likely to "await the fated onslaught, watchful but passive." Sims and Baumann attributed the disproportionately higher tornado death rate in the South to "fatalism, passivity, and perhaps most important, lack of trust in and inattention to society's organized system of warning."[51] Another study, by Richard J. Newcombe, focused on the reactions of individuals in states with varying frequencies of tornadoes. Newcombe found that residents of tornado-prone states such as Oklahoma responded more prudently to tornado warnings than did those who lived in states such as Pennsylvania, where tornadoes were less likely to occur.[52]

Perhaps John Grant Fuller best summarized the public's attitude toward tornado warnings in *Tornado Watch #211*, an account of a deadly tornado outbreak that struck eastern Ohio and western Pennsylvania on May 31, 1985, claiming seventy-six lives:

> Ostby [director of NSSFC] was concerned over two problems. There was the technical problem of trying to put out a fast and accurate forecast. There was the problem of

> convincing the public it ought to be ready—and to know what to do. He agreed with the sociologists. Many people will do almost anything they can to deny that they're under risk. If the warning is for Kansas, it can't happen in Missouri. If the neighbor down the street doesn't go to his basement, why should I?
>
> The syndrome was especially true in places where tornadoes rarely hit. In the Tornado Alley of the Great Plains, the audience was attentive to tornado and severe thunderstorm watches and warnings. They had faced tragedy too many times. In the East, the picture was different. Tornadoes were rare. But when they hit, they hit hard. The population is more dense. Indifference is widespread. The key thought: It can't happen here.[53]

Since the 1950s the NWS has spent billions of dollars to improve its tornado forecasting and warning systems and to educate the public about how to react when a tornado threatens their community. Automated data collection systems, satellite pictures, computers, and Doppler radar have aided severe weather forecasters in compiling and processing the massive amounts of data that flow into the SPC each day, but severe weather forecasters can only recognize that conditions are favorable for tornado formations within a designated area; they cannot foresee whether a tornado will actually materialize. Neither can they specify exactly where it will touch down. The best they can do is warn citizens in a high-risk area that tornadoes are possible and notify storm spotters to be on the watch for tornado development in their locale.

An extensive network of storm spotters and a nationwide Doppler radar network, along with substantial improvements in communications, especially television, have increased the chances a community will receive several minutes' warning of an approaching tornado. The ultimate goal is to ensure that every American receives adequate warning of an approaching tornado and takes shelter. To aid in achieving this goal the NWS has enlisted the schools and the media to educate the public on the meanings of tornado watches and warnings and the actions to take in each case.

CHAPTER 7

An Evaluation of the Integrated Tornado Warning System

The NWS integrated tornado warning system exists only for the purpose of saving lives. To judge the system's effectiveness in reducing tornado fatalities is difficult—no two tornadoes and no two communities are identical. No one knows how many Americans would have died in tornadoes during the last half century had no watch/warning system been instituted; therefore, no one can reasonably say that a certain number of lives were saved. Perhaps the best method of evaluation combines a look at the statistics with an examination of actual tornado occurrences to see how well each segment of the system functioned under "combat" conditions.

An examination of tornado statistics for the twentieth century reveals that tornado deaths have declined substantially since the 1950s (table 7). During that decade tornadoes killed 1,412 Americans. By comparison, tornado deaths numbered 509 for the 1980s and 577 for the 1990s. All of these numbers pale in comparison with the record 3,220 deaths during the 1920s. Although the substantial increase in tornado deaths during the 1970s would seem to negate the system's success, 307 of the fatalities occurred during the unprecedented April 3–4, 1974, outbreak.

TABLE 7

Tornado Deaths by Decade, 1900–1999

1900–1909	1839
1910–1919	1756
1920–1929	3169
1930–1939	1944
1940–1949	1786
1950–1959	1419
1960–1969	937
1970–1979	987
1980–1989	522
1990–1999	577

SOURCE: Statistics before 1990 from Thomas Grazulis, *Significant Tornadoes, 1680–1991*; statistics for 1990s from Storm Prediction Center, Historical Tornado Data Archive, http://www.spc.noaa.gov/archive/tornadoes.

Perhaps a more compelling statistic than total tornado deaths by decade is the number of deaths from a single tornado. Table 8 lists the twenty-five deadliest tornadoes in United States history. Only four of the twenty-five occurred after the institution of tornado watches in 1952, and all of them struck during the 1950s, before many communities had radar to spot the tornadoes or television to spread the warnings. Since the Udall, Kansas, tornado of May 25, 1955, only thirty tornadoes have killed more than twenty people each. Many of these occurred during the 1965 and 1974 outbreaks. From 1980 through 1997 only five tornadoes claimed more than twenty lives each.[1] The statistics are especially impressive when the population increase in the tornado-prone states is considered. From 1950 to 1990 the population in the Southeast and the Southern Plains increased almost 60 percent. Texas's population more than doubled, and Florida's increased fivefold.

TABLE 8

The 25 Deadliest U.S. Tornadoes

RANK	LOCATION	DATE	DEATHS
1	Tri-State (MO, IL, IN)	March 18, 1925	689
2	Natchez, MS	May 6, 1840	317
3	Saint Louis, MO	May 27, 1896	255
4	Tupelo, MS	April 5, 1936	216
5	Gainesville, GA	April 6, 1936	203
6	Woodward, OK	April 9, 1947	181
7	Amite, LA and Purvis, MS	April 24, 1908	143
8	New Richmond, WI	June 12, 1899	117
9	Flint, MI	June 8, 1953	115
10	Waco, TX	May 11, 1953	114
10	Goliad, TX	May 18, 1902	114
12	Omaha, NE	March 23, 1913	103
13	Mattoon, IL	May 26, 1917	101
14	Shinnston, WV	June 23, 1944	100
15	Marshfield, MO	April 18, 1880	99
16	Gainesville and Holland, GA	June 1, 1903	98
16	Poplar Bluff, MO	May 9, 1927	98
18	Snyder, OK	May 10, 1905	97
19	Natchez, MS	April 24, 1908	91
20	Worcester, MA	June 9, 1953	90
21	Starkville, MS and Waco, AL	April 20, 1920	88
22	Lorain and Sandusky, OH	June 28, 1923	85
23	Udall, KS	May 25, 1955	80
24	Saint Louis, MO	September 29, 1927	79
25	Louisville, KY	March 27, 1890	76

SOURCE: Storm Prediction Center, "The 25 Deadliest U.S. Tornadoes," http://www.spc.noaa.gov/archive/tornadoes/t-deadly.html.

A third statistic useful for judging the system's effectiveness is the average number of killer tornadoes per decade. Compared with the previous two sets of data, these statistics do not as readily demonstrate the effectiveness of the warning system. The average number of killer tornadoes (those that cause at least one death) per decade has not decreased substantially since the 1950s. In spite of the widespread use of Doppler radar and extensive television and radio warning systems, even the 1990s have not seen a notable decline in the number of killer tornadoes. What has declined, though, is the number of multiple-death tornadoes. From 1974 through 1996, 43.2 percent of all killer tornadoes claimed more than one victim. In comparison, from 1952 through 1973, 52.7 percent of all killer tornadoes took more than one life.[2] These statistics demonstrate that although radar and television warnings obviously cannot reduce the number or intensity of tornadoes, they have lowered the number of lives individual tornadoes take.

To evaluate its tornado watch/warning system's performance and to provide feedback to its forecasters, the NSSFC correlated tornado occurrences and death statistics with tornado watches and warnings. Joseph Galway studied the 497 killer tornadoes from 1952 through 1973 and found that 56 percent of these tornadoes and 66 percent of tornado deaths had occurred within the time and area of a tornado watch. Of considerable interest to Galway were the 217 deadly tornadoes that claimed 881 lives outside of watch areas. Galway noted that the distribution of tornado deaths outside a watch area was not uniform throughout the country (table 9): area 1 suffered 48 percent of these deaths. Although areas 2 and 3 had a comparable number of total tornado deaths, a much larger percent of the deaths occurred within a watch.[3] This discrepancy is particularly significant because Mississippi has the sad distinction of being second only to Texas in the number of tornado deaths from 1950 to 1995, and Arkansas and Mississippi top the list of tornado deaths per million people, killer tornadoes per

TABLE 9

Tornado Deaths for Selected Areas, 1952–1973

	No. of Deaths	% of National Deaths	No. of Deaths Outside Watch	% of Deaths Outside Watch
Area 1	777	30	421	48
Arkansas, Louisiana, Mississippi, Alabama, Georgia, Florida				
Area 2	703	27	115	13
Minnesota, Wisconsin, Illinois, Indiana, Ohio, Michigan				
Area 3	791	31	161	18
Texas, Oklahoma, Kansas, Nebraska, Iowa, and Missouri				
Totals	2271	88	697	79

SOURCE: Joseph Galway, "Relationship of Tornado Deaths to Severe Weather Watch Areas," *Monthly Weather Review* 103 (August 1975): 739.

million people, and killer tornadoes per ten thousand square miles.[4]

The first question these statistics bring to mind is why SELS/NSSFC issued so few tornado watches for the Southeast during this period, and the logical follow-up question is, how many lives would tornado watches in these areas have saved? A better question, though, is why did tornadoes take so many lives in areas 2 and 3 when tornado watches were in effect? It would appear that tornado watches by themselves did not substantially reduce the death toll. Preston Leftwich and John E. Hales of NSSFC conducted a comprehensive evaluation of the system's effectiveness from 1982 through 1988. Hales's study of significant tornadoes, those of F3 or greater intensity or those resulting in a death, found that 71 percent of these tornadoes for the period occurred in tornado watches. Eighty-four percent of catastrophic tornadoes, those of F4 or greater intensity or those resulting in four or more deaths, occurred in tornado watches during the same period. Unfortunately, a tornado warning preceded only 38 percent of the significant tornadoes and 31 percent of the catastrophic tornadoes. Like Galway, Leftwich found a great variation from state to state. Table 10 shows the percentage of tornadoes that occurred within a tornado watch for three representative states and the entire nation. These figures demonstrate Leftwich's observation that the citizens of the Southern Plains (Texas, Louisiana, Oklahoma, and Arkansas) had the best chance of having a tornado watch in place when a tornado occurred, whereas those who lived in the Northeast frequently had no advance notice of a tornado.[5]

Geographic variables such as the time of day and season of the year of tornado occurrences, as well as population density, spotter networks, and visibility (mountains or trees obscuring view), might explain the discrepancies in warnings from state to state, but they do not explain the variance in tornado watches, which were the sole responsibility of the NSSFC. The most obvious conclusions from the Galway and Leftwich studies are

TABLE 10

Yearly Percentages of Tornadoes within Watches/Warnings

Year	# of Tornadoes	Watch in Effect	Warning in Effect	Watch and Warning
		National		
1982	1088	42.5	28.6	18.2
1983	988	42.1	29.9	17.7
1984	1026	39.5	37.7	18.7
1985	767	40.9	30.0	16.9
1986	791	32.8	30.9	14.8
1987	695	25.3	24.0	8.3
1988	770	41.8	22.6	14.2
		Mississippi		
1982	21	23.8	23.8	4.8
1983	21	76.2	38.1	33.3
1984	21	52.4	33.3	19.0
1985	21	28.6	38.1	16.7
1986	48	37.5	18.8	12.5
1987	48	93.8	25.0	25.0
1988	67	74.6	40.3	35.8
		Pennsylvania		
1982	4	0.0	0.0	0.0
1983	20	20.0	20.0	10.0
1984	11	0.0	0.0	0.0
1985	49	57.1	20.4	10.2
1986	7	0.0	0.0	0.0
1987	6	0.0	0.0	0.0
1988	6	0.0	0.0	0.0
		Oklahoma		
1982	225	60.2	32.0	23.3
1983	188	61.9	55.4	42.4
1984	59	57.6	52.5	39.0
1985	37	43.2	43.2	27.0
1986	48	56.3	45.8	27.1
1987	23	26.1	21.7	4.3
1988	18	33.3	11.1	11.1

Source: Department of Commerce, NOAA, NSSL, Preston W. Leftwich and John E. Hales, *A Dyad of Papers Concerning Joint Verification of Severe Local Storm Watches and Warnings During Tornado Events,* NOAA Technical Memorandum NWS NSSFC-25, 8.

(1) the likelihood a tornado watch will be in effect when a tornado strikes vary according to locality, (2) many tornado fatalities occur outside tornado watch areas, and (3) other factors besides tornado watches affect the tornado death tolls. Since the mid-1990s, Doppler radar, advanced computers, remote satellite imaging, wind profilers, and increased research and understanding of the meteorological conditions that generate tornadoes increased the probability that the Storm Prediction Center would have a tornado watch in place when a killer tornado struck. For example, eighteen of the twenty-five tornado fatalities in 1996 occurred within a valid tornado watch, and four other deaths occurred close to a watch (within twenty-five miles or fifteen minutes). Only two of the sixty-nine tornado deaths in 1997 occurred outside a watch area.[6] These statistics demonstrate that the forecasting segment of the integrated tornado warning system has been improving, but it is only one part of the system.

Perhaps a better way to evaluate the system's effectiveness is to examine its operation during tornadoes that occurred in outbreaks, struck heavily populated areas, or were extremely deadly. Many factors, such as the time of day and year of the tornado's occurrence, the population density of the affected area, the type of housing, the availability of shelter, and the public's knowledge of actions to take, influence the death rate. Sociologists, psychologists, and engineers study tornado disasters to determine what role they can play in reducing fatalities. In a like manner, the NWS appoints a natural disaster survey team to evaluate how well its units functioned in the face of a deadly tornado. The survey team's role is to determine what the community did right, what went wrong, and what improvements are needed to avoid a future disaster.

Outbreaks, or the occurrence of six or more tornadoes spawned by the same weather system, severely test the tornado warning system. During outbreaks, tornadoes may form and disappear so quickly that affected communities may not receive advance warning of their approach. Frequently the storms down power and telephone lines, disrupting communications. Residents, bombarded by so many messages of impending danger, may become confused or simply ignore the warnings.

A major test of the tornado preparedness system came on Palm Sunday, April 11, 1965, when a massive outbreak of thirty-eight tornadoes, nineteen in the F4 or F5 category, tore through parts of six Midwest states, killing 256. Indiana alone suffered 137 fatalities.[7] Sociologist John Brouillette at Ohio State University's Disaster Research Center conducted an assessment of the tornado warning system in Indiana during the outbreak. The questions he asked have become the standard ones disaster researchers have asked since then: "Did the residents of these areas have any alert or warning of the tornadoes? If not, why not? If they were alerted or warned, why were the casualty figures so high?"[8]

Brouillette found that SELS had issued a tornado watch for southern Wisconsin, eastern Iowa, and northern Illinois at 1:00 P.M. and had extended the watch area to northern Indiana, southern Michigan, and northwestern Ohio at 4:20 P.M. The Kansas City unit had telephoned the watches to all of the local Weather Bureau Offices responsible for the areas, including South Bend, Fort Wayne, and Indianapolis in Indiana. The local bureau offices in turn had notified airports, spotter networks, state police and local law enforcement offices, and radio and television stations, which had passed on the information to the public. Although many stations were understaffed because of the holiday, a sufficient number of them had broadcast the tornado watch to allow the general public to take precautionary measures. Brouillette concluded that SELS and the local Weather Bureau Offices had given adequate forewarning of the pending disaster. If this statement is true, the question remains, why did so many die?

Brouillette placed the fault at the warning level. Downed telephone and power lines kept many from receiving warnings.

Some law enforcement agencies failed to pass on the warnings the Weather Bureau Offices sent them. Such was the case in one unidentified Indiana town when the chief of police, fearing the public might panic, did not notify the local radio station that a tornado was approaching. More commonly, the citizens did not hear the warnings because on the warm spring afternoon they were involved in activities other than watching television or listening to the radio. Most frightening to Weather Bureau personnel was the prevailing attitude that Brouillette found among many who heard the warnings: "It will never happen to my home." The study concluded that for lives to be saved outside of Tornado Alley the organizations involved in warning citizens had to continue repeating them to emphasize the seriousness of the situation.[9]

Another deadly outbreak struck the country on February 21, 1971, when an unknown number of tornadoes swept through rural areas of Mississippi and Louisiana, killing 119 and injuring more than two thousand. The three massive F4 tornadoes, which claimed all but two of the victims, were on the ground for an average of one hour and an average distance of 110 miles. The NSSFC had issued five different tornado watches for the area throughout the day, and all affected areas were in a valid watch. Lead times for tornado warnings the NWS offices in Shreveport, Memphis, and Jackson issued for the twenty-two affected counties averaged 0.9 hours, but five counties received no warnings. Table 11 shows the watch and warning lead times for the communities that suffered the worst losses. Radio and television stations saturated the Mississippi Delta area with the watches and warnings. Residents of the area told NWS disaster team members, "We heard the warnings—everyone knew they were coming." Why, then, did over one hundred people die? Actions the tornado survivors took indicated that they knew what precautions to take, but they encountered a major problem: lack of adequate shelter. Houses did not have basements; few had storm cellars. Many of the hard-struck communities did

TABLE 11

Tornado Watch and Warning Lead Times for Selected Communities

Community	Deaths	Watch Lead Time	Warning Lead Time
Cary, MS	14	2 hrs. 10 min.	50 min.
Delta City, MS	8	2 hrs. 10 min.	50 min.
Good Lake, MS	11	6 hrs. 30 min.	70 min.
Inverness, MS	19	6 hrs. 40 min.	25 min.
Little Yazoo, MS	9	3 hrs. 50 min.	95 min.
Pugh City, MS	22	6 hrs. 56 min.	89 min.
Madison Parish, LA	10	1 hr. 10 min.	0 min.

SOURCE: Department of Commerce, NOAA/NWS, *Mississippi Delta Tornadoes of February 21, 1971*, 7.

not have a substantial steel-reinforced building such as a school to serve as a shelter. Ditches were long distances from houses or filled with water. The houses in which most died were old and flimsily built, unable to withstand the storms. In spite of the poor shelter, the NWS estimated that the watch/warning service saved 350 lives.[10]

The worst tornado disaster in the United States after the institution of the watch/warning system occurred on April 3–4, 1974, when 148 tornadoes killed 315 people in eleven different states. One of the hardest-hit communities was Xenia, Ohio. Early on April 2, NSSFC and the Regional Warning Coordination Center at Fort Worth alerted all Weather Service offices and radar sites in the central and southern regions of the developing severe weather situation. Between 3:50 A.M. on April 3 and 2:00 P.M. on April 4, the NSSFC issued thirty watches. During the same period NWS offices issued more than 250 severe thunderstorm and tornado warnings. Most tornadoes occurred in or near valid watch areas, but several struck without warning. NWS personnel and the news media had difficulty keeping track of the unusually large number of watches and warnings. Several states had three or four overlapping tornado watches valid for the same time period. Nearly three-fourths of the warnings came between 2 and 8 P.M. Some tornadoes reached ground speeds of fifty to sixty miles per hour, which drastically reduced warning lead times. In spite of these problems, Weather Service offices were able to notify local broadcast media of the imminent dangers. In turn, the television and radio stations cooperated by interrupting regular programming with warnings. Although the death toll was very high, investigators believed that the dissemination of warnings and community actions saved thousands of lives.

Again, the question arises of why so many died. Part of the answer may lie with the residents of the affected areas. The public's willingness to recognize the danger and its knowledge of actions to take varied from one locale to the other. The disaster team's survey found that in those areas where the frequency of tornadoes was great, such as Alabama and Tennessee, people knew the difference between a watch and a warning and what actions to take in each instance. On the other hand, those in Kentucky, Ohio, and Indiana were often confused about the terminology or were oblivious to the threat a tornado could present. Many took shelter only in the last seconds when a tornado was bearing down upon them, while others remained in the open and survived only through good fortune.[11] Polk Laffoon, in his book on the Xenia tornado, wrote, "In southwestern Ohio, where Christmas is rarely white and Easter is usually unsunny, people have learned not to count on the weather but to cope with it the only way possible: they ignore it." Jay Bracken, disc jockey at Xenia's only radio station, related that in spite of the weather reports that tornadoes were moving into the area after devastating Indiana, he saw little cause for alarm and passed that thought on to his listening audience. After all, he had been in Xenia for seven years and had never given a tornado a second thought.[12] Unfortunately, an F5 tornado took thirty-four lives in Xenia, making this tornado the deadliest in the outbreak.

On November 21–23, 1992, another deadly outbreak struck the country. Ninety-five tornadoes raked parts of thirteen states from Texas to the mid–Atlantic coast, killing twenty-six and injuring more than six hundred. The NSSFC had identified the severe weather threat more than twenty-four hours before the first tornado touched down in southeast Texas and, over the course of the outbreak, issued sixteen tornado watches. Ninety-one of the tornadoes and twenty-five of the deaths occurred during a valid watch. Tornado warning statistics were not as impressive. Twelve of the twenty-six deaths occurred when no warnings were in effect. The NWS disaster team labeled the Houston office's performance exemplary and gave much credit to the newly installed WSR-88D (NEXRAD) radar for providing an average warning lead time of twenty-five minutes. Although five tornadoes, including one rated F4, crossed parts of heavily populated Harris County (Houston), no deaths occurred. Rankin County, Mississippi, was not as fortunate. No tornado warning was in effect when an F4 tornado moved through the county under cover of darkness and killed ten, including six in mobile homes and four in one brick home. The Jackson Weather Service Office did not have Doppler radar, and lacking any ground truth report from spotters, law enforcement, or the public, it did not issue a tornado warning.[13]

The tornadoes that struck Harris and Rankin Counties provided an excellent chance to evaluate the entire system. In both cases an F4 tornado struck during a valid tornado watch. Citizens of Houston's Channelview area, where the F4 tornado struck, were aware of the possibility of tornadoes from their own knowledge of local weather and from the radio and TV forecasts. Houston schools had trained their students from the third grade on what actions to take in tornadoes to save their lives. Neil Frank, retired director of the NWS National Hurricane Center and chief meteorologist at Houston television station KHOU, had encouraged the station to buy an automated system to generate severe weather crawls only a few months before the outbreak. He had been concerned that the three major network stations in Houston had not been warning the public in a timely manner and often had not even transmitted tornado warnings.[14] The citizens of Houston had everything the system could offer to save lives: a valid tornado watch, tornado warnings with adequate lead times, numerous television and radio stations with trained meteorologists, and a knowledge of actions to take.

In contrast, the citizens of Rankin County, Mississippi, had few of the advantages the Houston residents had. The tornadoes struck Houston during daylight hours but roared into the rural Mississippi county near midnight. A valid tornado watch was in effect, but a thirty-five-year-old weather radar set hindered the Jackson Weather Service Office from adequately accessing the storm's intensity and tornado-producing capability. The county's Emergency Management Agency was just getting organized at the time of the tornado, and its storm spotter network needed additional personnel and training. In spite of the fact that many Mississippi radio stations are daytime only and others reduce power at sunset, the NWS had not provided either radio or television stations with any warnings to broadcast.[15] How informed the Mississippi residents were about tornadoes and the precautions to take is unknown, but the assumption is that those who lived in that state, which has the highest tornado death rate both per capita and per square mile, would have some knowledge of the storms.

These examples illustrate that sometimes one segment of the integrated warning system is more at fault than the others. During the 1965 outbreak it was the communications or warnings systems that frequently failed, whereas in 1992 weaknesses in both the detection and warning segments may have caused several deaths in Mississippi. The public's lack of knowledge of how to respond to tornado warnings probably cost several lives during the 1974 outbreak, but in 1971 inadequate shelter doubtlessly contributed to the tornado toll in Mississippi.

Not all outbreaks produce a substantial number of fatalities. In fact, some never kill anyone. Table 12 lists outbreaks of forty or more tornadoes since the NWS's creation of a severe weather unit. With the exception of the massive 1965 and 1974 outbreaks, most claimed very few lives in comparison with the number of tornadoes. Only the May 15–16, 1968, outbreak, which killed thirty-three in Jonesboro, Arkansas, and the April 21, 1967, outbreak, which killed fifty-eight in the metropolitan Chicago area, claimed more than ten people at one location. One of the greatest success stories of the warning system occurred during the March 13, 1990, outbreak in Kansas. Two F5 tornadoes destroyed 120 homes in the town of Hesston, but the death toll was limited to one. The NWS, the broadcast media, and a siren system gave adequate warnings to the people of the community, who knew exactly what to do to escape death. The statistics would seem to indicate that with few exceptions the NWS tornado watch/warning program has been quite successful in preventing deaths during tornado outbreaks.

Although tornado outbreaks frequently overwhelm the warning system and individual tornadoes within an outbreak may sneak by without a warning, a catastrophic tornado (F4 or F5) presents a greater challenge to the weather community, especially if it threatens a heavily populated area. Only about 2 percent of all tornadoes are catastrophic, but they are responsible for more than 70 percent of all tornado-related deaths. Fortunately, many of these monsters have missed population centers, but enough have struck metropolitan areas to make those responsible for tornado warnings ever conscious of the damage they can inflict. Before the modern tornado warning system, cities such as Saint Louis, Louisville, Omaha, and Gainesville, Georgia, suffered massive losses from tornadoes.[16] In 1953 Waco, Flint, and Worcester made the headlines when tornadoes killed about one hundred in each city, but since that year cities have escaped the massive loss of life so common in earlier times. Watches and warnings may have saved hundreds

TABLE 12

Outbreaks with Forty or More Tornadoes, 1952–1996

Date	Total Tornadoes	F4/F5 Tornadoes	Killer Tornadoes	Total Deaths
April 3–4, 1974	148	30	49	315
Sept. 19–23, 1967	111	0	2	5
May 26–29, 1973	99	3	8	22
Nov. 21–23, 1992	95	5	9	26
May 18–19, 1995	80	2	2	4
May 11–12, 1982	70	1	1	2
April 26–27, 1994	67	1	1	3
March 20–21, 1976	66	3	3	3
June 2, 1990	66	7	4	9
June 16, 1992	65	2	1	1
June 26–27, 1994	62	0	2	2
May 2–3, 1984	60	1	1	5
March 13, 1990	59	4	2	2
June 15–16, 1992	58	2	0	0
June 8, 1993	58	0	0	0
May 8, 1988	57	0	0	0
April 26–27, 1991	53	5	5	21
May 25–26, 1965	51	0	0	0
May 4–5, 1959	49	0	0	0
April 11–12, 1965	48	19	21	256
April 2–3, 1982	48	4	10	29
May 18–20, 1983	48	0	6	6
Jan. 9–10, 1975	47	1	3	11
May 15–16, 1968	46	4	8	74
April 7–9, 1980	46	0	1	1
May 16, 1991	46	0	0	0
May 7, 1993	46	1	1	1
April 21, 1967	45	4	3	58
June 7–8, 1984	45	2	4	13
Nov. 15, 1988	44	0	3	7
April 29, 1984	42	1	1	1
May 29, 1980	40	0	0	0
Dec. 14, 1971	40	0	2	2
Nov. 7, 1995	40	0	0	0

Source: Grazulis, *Significant Tornadoes, 1680–1995*, 1396.

of lives in Kansas City and Topeka when tornadoes devastated the cities in 1957 and 1966, respectively.

Lubbock, Texas, is another city in which tornado warnings may have saved many lives. An F5 tornado, one-and-one-half miles wide in places, plowed an eight-and-one-half-mile long path through the downtown and residential areas of the South Plains city of 150,000 on May 11, 1970. Although the NSSFC had not issued a tornado watch for the area, developing thunderstorms alerted the Lubbock Weather Bureau Office to the possibility of severe weather. Meteorologists at the Lubbock office tracked the thunderstorms on their older WSR-1 radar, and their counterparts in Amarillo aided them with their newer WSR-57 radar, which could better determine cloud top heights. When the Lubbock radar displayed a telltale hook echo about seven miles from the airport and a spotter reported a funnel cloud and large hail, the Weather Bureau Office issued a tornado warning for the city at 8:15 P.M., but the funnel cloud never touched the ground. During the next one-and-one-half hours other hook echoes appeared on the radar screen, and shortly after 9:35 the violent tornado struck. Radio and television stations had been cutting into regular programming to broadcast warnings and give information on precautions to take since the initial warning, but in spite of the advance notice, the tornado killed twenty-six and injured fifteen hundred. About half of the deaths occurred among the Spanish-speaking population, perhaps because the city's only Spanish-language radio station had gone off the air at 8 P.M. Doubtless, an extensive education program conducted by the school system, *Avalanche-Journal*, radio and television stations, and the Weather Bureau saved lives as citizens huddled under beds and tables and pulled mattresses over their heads to shelter themselves from the storm's fury. The Weather Bureau survey team conservatively estimated that the warnings and resulting actions of an informed citizenry had saved at least 125 lives.[17]

Another city familiar with tornadoes is Omaha, Nebraska. The 1913 tornado killed ninety-five, and six smaller tornadoes had caused no deaths when they struck the city. Citizens of Omaha might have been more attentive than usual to the NSSFC's tornado watch on May 6, 1975, because a weak tornado had caused some property damage and injuries only two months earlier. About forty-five members of REACT (Radio Emergency Associated Communications Team), an organization of amateur radio operators trained as spotters, moved to their assigned vantage points, where they might sight a tornado before it reached the city. When REACT spotted a tornado near the city, the NWS issued a tornado warning for Omaha at 4:14 P.M. Civil Defense sirens, local radio and television stations, and NOAA Weather Radio alerted Omaha residents, who sought shelter in closets, basements, and under beds just as the media, which had been conducting an extensive educational campaign since the March tornado, had instructed them to do. At 4:29 an F4 tornado began churning its way through the city, striking homes, apartment buildings, businesses, churches, schools, and a hospital. Officials estimated that thirty thousand people lived or worked in the affected area. After the funnel withdrew into the clouds, the city began assessing the damage. The findings were staggering: The $400 million price tag made it the costliest tornado in American history. Fortunately, the death toll did not match the damage toll—three died and three hundred suffered injury. Omaha public safety director Richard Roth said, "The toll could have been anywhere from 300 to 500 on up," and officials credited a fifteen-minute warning and an educated public with keeping the fatalities to a minimum.

Whenever an F4 or greater tornado strikes a major metropolitan area, the results can be catastrophic. Fortunately, this was not the result when a three-hundred-yard-wide F4 tornado tore through Lancaster, Texas, a Dallas suburb of twenty-two thousand people, on April 25, 1994. Although the storm destroyed or severely damaged over two hundred homes, demolished the historic Lancaster Town Square, and caused over $200 million in damages, the death toll was three. The low death toll was the result of all parts of the tornado watch/warning system working perfectly. Early in the day the NSSFC had issued a Public Severe Weather Outlook to emphasize the volatility of the situation and, at 10:56 A.M., had placed the first of three tornado watches in effect. When thunderstorms formed near Fort Worth in the late afternoon, the 650-member Dallas/Fort Worth Metroplex SKYWARN spotter network went into action. At the Fort Worth Weather Service Field Office meteorologists tracked the storms on their $2.5 million, three-month-old WSR-88D and throughout the evening issued eight tornado warnings based on Doppler radar or storm-spotter reports. Radio and television stations, some with their own Doppler radars, rapidly passed the warnings to the public. Sirens throughout the Metroplex, including ten in Lancaster, wailed to notify the citizens to turn on their televisions for further information. The residents of Lancaster knew what to do when tornadoes threatened and accordingly took cover in hallways and interior rooms. At a fast-food restaurant, employees hurried customers into the meat locker, where they rode out the storm uninjured. The *Dallas Morning News* and the *Fort Worth Star-Telegram* praised the Doppler radar, and the timely warnings it enabled the Fort Worth office to issue, for saving lives.[19]

Not every community has been as fortunate as Lancaster and Omaha. Thirty died when an F4 tornado devastated the tiny southwest Texas community of Saragosa on May 22, 1987. Although a severe thunderstorm watch and a tornado warning were in effect, many residents of Saragosa were attending a Head Start graduation ceremony at the community center and did not hear the warnings over television or radio. Only a few minutes before the tornado struck, someone ran in to warn of the tornado's approach, but those in the cinder-block building had nowhere to take shelter except in the building itself. Parents put the children under the available tables and shielded them with their own bodies, but when the building's walls collapsed, twenty-two died.[20]

One of the most-studied tornadoes of recent years has been the F5 monster that devastated Andover, Kansas, on April 26, 1991. That day fifty-four tornadoes swept across parts of six states, but only the Andover storm reached F5 magnitude. As is usual with an outbreak, the NSSFC had issued a tornado watch six hours in advance. When Wichita's storm-spotter network—one of the oldest and best trained in the country—spotted the tornado, the Wichita Weather Service Office transmitted the warning to all radio and television stations. The tornado, with an intensity of F2 and F3, tracked through McConnell Air Force Base, causing extensive damage and missing by only a thousand feet the flight line, where eighty-four military planes, including two B-1B bombers containing nuclear warheads, were parked. Along its path to Andover the tornado intensified to an F5. The residents of Andover had a ten-minute warning. Directly in the tornado's path was the Golden Spur Mobile Home Park, which contained 241 mobile homes and approximately seven hundred residents. Before the storm struck, about half of those who had been at home fled the park, and 149 sought refuge in the park's community shelter. Thirty-eight remained in their trailers, and eleven of them perished when the tornado leveled all but eight of the mobile homes.[21]

Whether or not they received them or acted upon them, the residents of Andover and Saragosa did have tornado warnings. The people of Plainfield and Crest Hill, Illinois, were not as fortunate. A multiple-vortex F5 tornado killed twenty-nine in the communities twenty-five miles southwest of Chicago about 3:15 P.M. on August 28, 1990, when it leveled three hundred homes, several shopping centers, apartment complexes, and schools. The NSSFC had issued a severe thunderstorm watch for northern Illinois until 8:00 P.M., but in spite of the appearance of supercells and reports of tornadic activity in the area just to the west of Plainfield, the staff did not upgrade to a tornado watch. The affected counties did not activate their spotters; Will County (Plainfield) did not even have an organized spotter effort but relied on law enforcement agencies. Radar at the Weather Service Meteorological Observatory at Marseilles, about thirty miles west of Plainfield, displayed the mesocyclone signature associated with a supercell thunderstorm or a tornado, but neither the staff at Marseilles nor the Chicago Weather Service Office, which had the responsibility for Plainfield, recognized the danger. As a result, the Chicago office did not issue any warnings for the massive tornado bearing down on the unsuspecting town. The broadcast media placed little emphasis on severe thunderstorm watches and warnings, which were common occurrences in their part of the country. However, they reported to the NWS's committee investigating the disaster that they would have given much more attention to a tornado warning. In spite of lack of warning, citizens aware of the approaching tornado took shelter in interior bathrooms, just as they had learned during tornado safety week activities or in the local schools' extensive tornado safety drills. Although the tornado was devastating by any standards, its legacy could have been much worse. About fifteen hundred students would have been inside the flattened school buildings had the tornado occurred the following day, the first day of school.[22]

What can be concluded from this examination of a variety of violent and deadly tornadoes? The reasons tornado deaths continue to occur, in spite of a well-developed, integrated warning system, are numerous. Some fault is within the system, such as failure to issue a tornado watch, inadequate detection (although with the widespread use of Doppler radar this is becoming less of a problem), or lack of tornado warnings that give the public time to take action. Some responsibility belongs to the public for failing to heed the issued watches and warnings and for not adequately providing for shelter before the storms appear. But sometimes, as with the Jarrell tornado, no fault can be assigned.

Some suggest that improved building construction, along with random chance, played a major role in the reduction of tornado deaths in the U.S. during the last half of the twentieth century. These factors indeed may have contributed to the decline in fatalities, but accounts of violent tornadoes that took few lives, such as the ones that struck Hesston, Kansas, in 1990 and Lancaster, Texas, in 1994, indicate that warnings and education are the key factors in saving lives. Although the ultimate goal of the integrated tornado warning system, preventing loss of life from tornadoes, becomes less achievable as Americans continue to move into tornado-prone areas of the country, improvements in tornado forecasting and the tornado warning system and public education should continue to lower the number of tornado fatalities.

NOTES

CHAPTER 1

1. G. Terence Meaden, "Tornadoes and Tornado Worship in Britain between 5500 and 4000 Years Ago," *Journal of Meteorology* 21 (May/June 1996): 187–97.

2. Gretchen Will Mayo, *North American Indian Tales: Earthmaker's Tales*, 23–26.

3. G. B. Bathurst, "The Earliest Recorded Tornado?" *Weather* 19 (July 1964): 202–4.

4. Aristotle, *Meteorologica*, trans. by H. D. P. Lee, 3.237.

5. Pliny the Elder, *The Historie of the World*, trans. by Philemon Holland, 1:24–25.

6. One of the few accounts of a European tornado to appear in the eighteenth century was Father Ruggiero Boschovich's account of a tornado that struck Rome in 1749. The English summary of the Jesuit priest's account appears in the December 1750 London *Monthly Review*.

7. John Winthrop, *History of New England, 1630–1649*, 2, 126.

8. Increase Mather, *An Essay for the Recording of Illustrious Providences; Wherein an account is given of many remarkable and very memorable events, which have hapned this last age, especially in New England*, 313–16.

9. David M. Ludlum, *Early American Tornadoes, 1586-1870*, 201–8. By comparison Thomas Grazulis lists seventeen tornadoes for the same period in *Significant Tornadoes, 1660–1991*, 6. This latter book contains hundreds of tables and maps and describes every significant tornado (one that caused death or had wind speeds exceeding 112 miles per hour). A supplement updates the information through 1995.

10. John Perkins, "Conjectures Concerning Wind and Water-spouts, Tornadoes, and Hurricanes," *Transactions of the American Philosophical Society* 2 (1786): 342.

11. Nathan G. Goodman, ed., *A Benjamin Franklin Reader*, 451–53; 389.

12. Benjamin Franklin, "Physical and Meteorological Observations, Conjectures, and Suppositions," *Philosophical Transactions* 55 (1765): 191.

13. Robert Hare, "On the Causes of the Tornado, or Water Spout," *American Journal of Science and Arts* 32 (July 1837): 154–56; "Notices of Tornadoes, Etc.," *American Journal of Science* 38 (1840): 80–81.

14. William C. Redfield, "Remarks Relating to the Tornado Which Visited New Brunswick, in the State of New Jersey, June 19, 1835, with a Plan and Schedule of the Prostrations Observed on a Section of Its Track," *Journal of the Franklin Institute* 32 (July 1841): 41.

15. Ludlum, *Early American Tornadoes*, 158–77.

16. James Pollard Espy, "Theory of Rain, Hail, Snow and the Water Spout, Deduced from the Latent Caloric of Vapour and the Specific Caloric of Atmospheric Air," *Transactions of the Geological Society of Pennsylvania* 1 (1835): 342–46. This is the basic principle of the convective theory of rainfall that is generally accepted today. Although Espy did not specifically mention tornadoes, scientists of the day considered these "landspouts" and waterspouts to be the same phenomenon.

17. Alexander Bache, "Notes and Diagrams, Illustrative of the Directions of the Forces Acting at and near the Surface of the Earth, in Different Parts of the Brunswick Tornado of June 19, 1835," *American Philosophical Society Transactions*, n.s., 5 (1837): 417.

18. James Rodger Fleming, *Meteorology in America 1800–1870*, 31.

19. William C. Redfield, "Remarks on the Prevailing Storms of the Atlantic Coast of the North American States," *American Journal of Science* 20 (July 1831): 17.

20. William C. Redfield, "On the Courses of Hurricanes, with Notices of the Typhoons of the China Sea, and Other Storms," *American Journal of Science* 35 (January 1839): 201.

21. Redfield, "Remarks Relating to the Tornado," 41.

22. William Ferrel, "The Motions of Fluids and Solids Relative to the Earth's Surface," *American Journal of Science* 31 (January 1861): 42; *Treatise on Winds*, 348.

23. Theodore Wiseman, *Origin and Laws Governing Tornadoes, Cyclones, Thunderstorms, and Kansas Twisters* , 5–13. According to Grazulis in *Significant Tornadoes* the actual number of tornadoes in Kansas during this period was five in 1880, eleven in 1881, eleven in 1882, twelve in 1883, and three in 1884.

24. John P. Finley, "Tornadoes: First Prize Essay," *American Meteorological Journal* 7 (August 1890): 166. For a more complete account

of Finley's role in tornado history see chapter 2.

25. Alexander McAdie, "Tornadoes, Second Prize Essay," *American Meteorological Journal* 7 (August 1890): 187.

26. Henry A. Hazen, "Tornadoes: A Prize Essay," *American Meteorological Journal* 7 (August 1890): 217.

27. Edward M. Brooks, "Tornadoes and Related Phenomena," in *Compendium of Meteorology*, ed. Thomas F. Malone, 676.

28. See chapter 2 for further explanation of the ban.

29. Alfred A. Henry et al., *Weather Forecasting in the United States , 67; William J. Humphreys, "The Tornado and Its Cause," Monthly Weather Review* (hereafter *MWR*) 48 (April 1920): 212. Humphreys was not one of the authors of *Weather Forecasting in the United States*, so there is no logical explanation for his taking credit for the article in the *MWR*. Humphreys identified himself as a "meteorological physicist" on the title page of his books.

30. W. J. Humphreys, "The Tornado," *MWR* 54 (December 1926): 501–3; *Physics of the Air*.

31. Patrick Hughes, *A Century of Weather Service, 1870–1970: A History of the Birth and Growth of the National Weather Service* , 61; Robert Marc Friedman, *Appropriating the Weather: Vilhelm Bjerknes and the Construction of a Modern Meteorology*, 161–63.

32. J. R. Lloyd, "The Development and Trajectories of Tornadoes," *MWR* 70 (April 1942): 73–74. Months of peak tornado occurrences do vary by region of the United States. The prime season is March through May in the Southeast. April through June are the most active months in the Midwest, the Northwest, and the southern Great Plains. The northern Great Plains, the Rocky Mountain region, and New England experience the most tornadoes June through August. The Gulf Coast states have a second tornado season in November; see Department of Commerce, National Oceanic and Atmospheric Administration, National Weather Service (herafter NOAA/NWS), *Tornadoes, Nature's Most Violent Storms: Preparedness Guide*, 5.

33. Morris Tepper, "On the Origin of Tornadoes," *Bulletin of the American Meteorological Society* (hereafter cited as *Bull AMS*) 31 (November 1950): 311–14; "The Tornado and Severe Storm Project," *Weatherwise* 4 (June 1951): 51–53; *Pressure Jump- Lines in Midwestern United States*.

34. D. Lee Harris, "Effects of Atomic Explosions on the Frequency of Tornadoes in the United States," *MWR* 82 (December 1954): 369. Harris reports 532 tornadoes in 1953, but official statistics from the National

Climatological Center list only 421 tornadoes for the year. Previous to 1953 the largest number of tornadoes was 262 in 1951. The death toll for 1953 was 515, one of the largest in the country's history. On May 11 a tornado killed 114 in Waco, Texas. A June 8 twister killed 115 in Flint, Michigan, and the following day a tornado killed ninety-four in Worcester, Massachusetts.

35. Edward M. Brooks, "The Tornado Cyclone," *Weatherwise* 2 (April 1949): 32; "Tornadoes and Related Phenomena," 677.

36. K. A. Browning, "The Organization of Severe Local Storms," *Weather* 23 (October 1968): 429–34.

37. *Stormwatch*, prod. Martin Lisius, 30 min., Texas Severe Storms Association, 1995, videocassette.

38. Japanese-born Tetsuya Theodore Fujita, known as Mr. Tornado in the weather community, taught and conducted extensive severe storm research at the University of Chicago from 1956 to 1990. He developed an interest in damage patterns after ground-truth expeditions to Hiroshima and Nagasaki in September 1945 and conducted his first tornado survey in Enoura, Japan, in 1948. See Tetsuya Theodore Fujita, *The Mystery of Severe Storms*, 181, 194.

39. Richard Bedard, *In the Shadow of the Tornado*, 107; Fujita, *Severe Storms*, 35.

40. Grazulis, *Significant Tornadoes*, 141–44.

41. John D. Isaacs et al., "Effect of Vorticity Pollution by Motor Vehicles on Tornadoes," *Nature* 253 (January 24, 1975): 254–55. According to the data in the article, from 1950 to 1973 only 1868 tornadoes occurred on Saturday whereas the daily average was 2176.

42. "Tornado Traffic," *Science Digest* 77 (May 1975): 28–29.

43. Howard Bluestein and Joseph Golden, "A Review of Tornado Observations," in *The Tornado: Its Structure, Dynamics, Prediction, and Hazards*, ed. Christopher Church et al., 330–31. TOTO resides at NOAA Headquarters in Washington, D.C.

44. Robert Davies-Jones, "Tornadoes," *Scientific American* 273 (August 1995): 56–57.

45. *Tornado Video Classics III*, prod. Thomas Grazulis, 120 min., Tornado Project, 1995, videocassette; Eric Rasmussen, "Tornadoes Targeted in VORTEX," *NSSL Briefings* (Summer 1995).

46. Eric Rasmussen, "New Findings on the Origins of Tornadoes from VORTEX," *NSSL Briefings* (Summer/Fall 1998).

47. Joshua Wurman, "Radar Observations of Tornadoes and Thunderstorms Experiment," in the Doppler on Wheels homepage

[online] [cited March 24, 1999], available from http://aaron.ou.edu/dow/index.html.

CHAPTER 2

1. Robert V. Bruce, *The Launching of Modern American Science, 1846–1876*, 194; Hughes, *Century of Weather Service*, 5; Gary K. Grice, ed., *The Beginning of the National Weather Service, the Signal Service Years (1870–1891)*, 1.

2. Ludlum, *Early American Tornadoes*, 156; "Queries Relative to Tornadoes," *Smithsonian Miscellaneous Collection* 10 (1873): 1–4.

3. Truman Abbe, *Professor Abbe and the Isobars: The Story of Cleveland Abbe, America's First Weatherman*, 100–101.

4. *U.S. Statutes at Large* (1870): 369; Donald R. Whitnah, "The United States Weather Bureau: Its Scientific Development and Public Services, 1870–1941," (Ph.D. diss., University of Illinois, 1957), 43–46.

5. The official United States weather organization has changed names and departments several times. From 1870 to 1890 the Division of Telegrams and Reports for the Benefit of Commerce under jurisdiction of the Army Signal Corps was in charge of weather services. In 1891 the weather agency came under civilian control. The U.S. Weather Bureau was part of the Department of Agriculture from 1891 to 1940. The Department of Commerce assumed control in 1940, and in 1970 the name became the National Weather Service.

6. Editors of the Army Times, *A History of the U.S. Signal Corps*, 57.

7. Gary Grice, "Evolution to the Signal Service Years (1600–1891): Personal View of Professor Cleveland Abbe," [online], [cited 7 February 1997], available from http://205.156.54.206/pa/history/abbe.htm.

8. Grice, *Beginning of the National Weather Service*, 2, 4.

9. Department of Commerce, NOAA/NSSL, *J. P. Finley: The First Severe Storms Forecaster*, by Joseph G. Galway (hereafter Galwey, *Finley*), NOAA Technical Memorandum ERL NSSL-97, 1–2.

10. Gary Grice, "Evolution to the Signal Service Years (1600–1891): Personal View of John P. Finley," [online], [cited February 7, 1997], available from http://205.156.54.206/pa/history/finley.htm; William Blasius, *Storms: Their Nature, Classification and Laws*.

11. John P. Finley, "Tornado Study—Its Past, Present, and Future," *Journal of the Franklin Institute* 71 (April 1886): 241–62; Galway, *Finley*, 5.

12. Finley, "Tornado Study," 254.

13. John P. Finley, "Intelligence from American Scientific Stations," *Science* 3 (1884): 767–68.

14. Finley, "Tornado Study," 255–56.

15. John P. Finley, "Tornado Predictions," *American Meteorological Journal* 1 (July 1884): 85–88.

16. G. K. Gilbert, "Finley's Tornado Predictions," *American Meteorological Journal* 1 (September 1884): 166–72.

17. G. E. Curtis, "Tornado Predictions and Their Verification," *American Meteorological Journal* 4 (1887): 68–74.

18. For a lengthy discussion of this topic see Allan H. Murphy, "The Finley Affair: A Signal Event in the History of Forecast Verification," *Weather and Forecasting* 11 (March 1996): 3–20.

19. Gustavus Hinrichs, "Tornadoes and Derechos," *American Meteorological Journal* 5 (November 1888): 313, (January 1889): 392.

20. William A. Eddy, "Letter to the Editor," *Scientific American* 53 (October 31, 1885): 277.

21. Army Signal Corps, *Report of the Chief Signal Officer of the Army*, 1885, 52; *Report of the Chief Signal Officer of the Army*, 1886.

22. For a detailed account of the Allison Commission see Whitnah, "United States Weather Bureau," 91–100.

23. Galway, *Finley*, 8–10.

24. U.S. Army Signal Corps, *Report of the Chief Signal Officer of the Army*, 1887, 21–22.

25. John P. Finley, *Tornadoes: What They Are and How to Observe Them; with Practical Suggestions for the Protection of Life and Property*, 25–33.

26. Until the late 1960s the Weather Service urged citizens to heed this advice and take shelter in the corner of the basement or first floor of a structure toward the storm (west, south, or southwest). After a study of the destruction patterns of the 1966 Topeka tornado, Joe Eagleman determined these were the worst places to take shelter and that the safest locations were those opposite the storm's approach or in the center of the structure. As a result, the Weather Service changed its policy and began instructing the public to take shelter in the interior section of the house or basement. See J. R. Eagleman, V. U. Muirhead, and N. Willems, *Thunderstorms, Tornadoes, and Damage to Buildings*.

27. Finley, *Tornadoes: What They Are*, 43.

28. Edward S. Holden, "A System of Local Warning Against Tornadoes," *Science* 37 (October 19, 1883): 521–22.

29. Army Signal Corps, *Report of the Chief Signal Officer of the Army*, 1890, 34–35.

30. Henry A. Hazen, *The Tornado*, 124–25.

31. *U.S. Statutes at Large* (1890): 653.

32. Joseph G. Galway, "Early Severe Thunderstorm Forecasting and Research by the United States Weather Bureau," *Weather and Forecasting*, 7 (December 1992), 568.

33. Department of Agriculture, Weather Bureau, *Report of the Chief of the Weather Bureau, 1895–1896*, xiii–xxv.

34. Cleveland Abbe, "No Increase in Tornadoes," *MWR* 27 (April 1899): 158.

35. Cleveland Abbe, "The Prediction of Tornadoes and Thunderstorms," *MWR* 27 (April 1899): 159–60.

36. During 1899 the southern Great Plains witnessed very few tornadoes. Oklahoma reported only three significant tornadoes (F2 or greater) for the year, all of which occurred after Widmeyer's May article, while Texas reported only one. Strangely, Kansas reported no significant tornadoes for the entire year (Grazulis, *Significant Tornadoes*).

37. Cleveland Abbe, "Unnecessary Tornado Alarms," *MWR* 27 (June 1899): 255.

38. Mark W. Harrington, *About the Weather*, 164–65. Because the annual number of tornadoes is much higher than Harrington knew about (approximately one thousand), the odds would be much less than he calculated, perhaps 1 in 300 in the 1990s.

39. Fred C. Bates, "Severe Local Storm Forecasts and Warnings and the General Public," *Bull. AMS* 43 (July 1962): 288–91.

40. Senate, *Agriculture Department Rules and Regulations*, 59th Cong., 2nd sess., 1906, S. Doc 298, serial 5091.

41. In the 1920s tornadoes killed 2,171 Americans, including 794 in 1925, the highest yearly total in American history. During the 1930s 1,944 perished in tornadoes in the United States. From National Oceanic and Atmospheric Administration, National Climatic Data Center, *Storm Data*, 1995, 40.

42. Marie Hall, interview by author, Lubbock, Texas, December 9, 1996.

43. Gary Lockhart, *The Weather Companion: An Album of Meteorological History, Science, Legend, and Folklore*, 148.

44. B. M. Varney, "Aerological Evidence as the Causes of Tornadoes," *MWR* 54 (April 1926): 163–65.

45. E. Durand-Gréville, "Squalls and the Prediction of Tornadoes," *MWR* 42 (February 1914): 97–99.

46. Peter S. Felknor, *The Tri-State Tornado: The Story of America's Greatest Tornado Disaster*, 6–8.

CHAPTER 3

1. Hughes, *Century of Weather Service*, 48–50; *U.S. Statutes at Large* 44 (1926):568.

2. *U.S. Statutes at Large* 52 (1938):1014; 54 (1940):1236.

3. Hughes, *Century of Weather Service*, 82.

4. Department of Commerce, Weather Bureau, *World War II History of the Department of Commerce, Part 10: U.S. Weather Bureau*, 6.

5. Office of Censorship, *A Report on the Office of Censorship*, 36.

6. Hughes, *Century of Weather Service*, 110.

7. F. W. Reichelderfer to all Weather Bureau stations, December 6, 1940, Records of the Weather Bureau, Records of the Office of the Chief, Office Files of F. W. Reichelderfer, Personal Subject Files 1939-63, National Archives II, College Park, Maryland, RG 27, Box 4.

8. Department of Commerce, *World War II History*, 32.

9. Ibid., 32–34.

10. Department of Commerce, Weather Bureau, *Organization and Operation of the Severe Storm Warning Service*, 14–15.

11. Bureau of Aeronautics, Navy Department, *Tornadoes*, 7.

12. Ibid., 1–2.

13. J. A. Riley to People of Tornado Warning Areas, 5 April 1944, Records of the Weather Bureau, Office of the Regional Director, Kansas City, Missouri, National Archives Branch Depository, Kansas City, Missouri, RG 27, Box 278 (hereafter cited as Regional Director Archives).

14. Francis W. Reichelderfer pioneered in aviation meteorology and served in the navy's weather service from 1922 to 1939. In this capacity he was responsible for the military's adoption of the Norwegian school's (Bjerknes) front theories and meteorological methods. He served as chief of the Weather Bureau from 1939 to 1963.

15. Dept. of Commerce, *Severe Storm Warning Service*, 16.

16. H. M. Van Auken to Victor Phillips, July 19, 1948, Regional Director Archives, Box 278.

17. Victor Phillips to J. A. Riley, July 21, 1948, Regional Director Archives, Box 278.

18. The other tornadoes in Sedgwick County for this period were on June 8, 1941, when one died, and June 20, 1942, when no fatalities occurred; Grazulis, *Significant Tornadoes*, 301–6, 939.

19. Memorandum for SR&F, May 3, 1949, Records of the Weather Bureau, Records of the Office of the Chief, Office Files of F. W. Reichelderfer, General Correspondence 1939-63, National Archives II, College Park, Maryland, RG 27, Box 10 (hereafter cited as Reichelderfer Archives).

20. Department of Commerce, Weather Bureau, *Weather Bureau Topics and Personnel* (hereafter *Weather Bureau Topics*), 9 (January 1950): 57.

21. J. A. Riley to Regional Weather Bureau Offices, August 6, 1942, Regional Director Archives, Box 1.

22. Department of Commerce, Weather Bureau, *Circular Letters*, 1943, at NOAA Central Library, Silver Spring, Maryland.

23. F. W. Reichelderfer to Fort Worth Regional Director, May 3, 1949, Reichelderfer Archives, Box 10.

24. Department of Defense, Air Force, Air Weather Service, *Weather Service Bulletin No. 4*, 24.

25. John Lewis, Charlie Crisp, and Robert Maddox, "Colonel Robert C. Miller and His Approach to Severe Weather Forecasting," an unpublished manuscript, April 1996, in author's personal collection.

26. Only twenty-six significant tornadoes occurred in California from 1880 through 1995. No tornado deaths have ever occurred in California; Grazulis, *Significant Tornadoes*, 249.

27. The Air Weather Service in the postwar period employed this radar, originally used as a bomb-aiming radar on B-29s in World War II, in storm detection. It had a range of one hundred miles.

28. Robert C. Miller, "The Unfriendly Sky," unpublished manuscript, in author's possession (hereafter cited as Miller manuscript). Statistics of damage are from *Weather Service Bulletin No. 4*, 24.

29. Lloyd, *Development and Trajectories*, 65–75; Department of Commerce, A. K. Showalter and J. R. Fulks, *Preliminary Report on Tornadoes*.

30. Ernest J. Fawbush, Robert C. Miller, and L. G. Starrett, "An Empirical Method of Forecasting Tornado Development," *Bull. AMS* 32 (January 1951): 5–6.

31. Ibid.

32. Miller manuscript.

33. Galway, "Early Severe Thunderstorm Forecasting," 576.

34. Miller manuscript; statistics from Grazulis, *Significant Tornadoes*, 943.

35. U.S. Weather Bureau Office, Amarillo, Texas, Station Log, 15 May 1949.

36. *Amarillo Globe*, May 16, 1949.

37. F. W. Reichelderfer to Senator Lyndon Johnson, May 27, 1949, Reichelderfer Archives, Box 10.

38. F. W. Reichelderfer to All District Forecast Centers, June 6, 1949, Reichelderfer Archives, Box 10.

39. Miller manuscript.

40. Ibid.

41. Ernest J. Fawbush and Robert C. Miller, "Forecasting Tornadoes," *Air University Quarterly Review* 6 (Spring 1953): 108–17.

42. Fawbush et al., "Empirical Method," 8–9.

43. Department of Defense, Air Force, Air Weather Service, *Tornadoes and Related Severe Weather*, 3.

44. Lyndon Johnson to Thomas K. Finletter, January 10, 1953, Office Files of Gerald W. Siegel, Preparedness Subcommittee Correspondence III, Papers of Lyndon B. Johnson, U.S. Senate 1949–61, Committee on Armed Services, Lyndon B. Johnson Library, Austin, Texas. Meteorologists did not note the correlation between a tornado and its radar signature, a hook echo, until 1953. Before then radar could detect a thunderstorm, but operators had no indication whether or not the storm contained a tornado.

45. "Sudden Attack," *Time*, September 22, 1952, 27; "SAC Crippled," *Aviation Week*, September 22, 1952, 18.

46. "Weather Officers Commended," *Take-Off* , January 16, 1953. This is the Tinker Air Force Base, Oklahoma, newspaper.

CHAPTER 4

1. Reichelderfer to All First Order Weather Bureau Stations, July 12, 1950, Reichelderfer Archives, Box 14.

2. *Miami Herald*, March 7, 1952.

3. F. W. Reichelderfer to Roy Calvin, March 6, 1952, Reichelderfer Archives, Box 14.

4. Ken Miller radio broadcast over station KVOO, Tulsa, Oklahoma about March 3, 1952, a transcript typed from a recording, Reichelderfer Archives, Box 14.

5. F. W. Reichelderfer Telephone Conversation with Erle Hardy, March 29, 1952, a transcript, Reichelderfer Archives, Box 14.

6. F. W. Reichelderfer Telephone Conversation with Erle Hardy from Oklahoma City, Okla., 10:00–11:00 A.M. March 26, 1952, a transcript, Reichelderfer Archives, Box 14.

7. F. W. Reichelderfer Telephone Conversation with Colonel Thomas S. Moorman, a transcript, March 13, 1952, Reichelderfer Archives, Box 14.

8. F. W. Reichelderfer to Brotzman, May 14, 1952, Reichelderfer Archives, Box 14.

9. For Project Leaders Conference, April 2, 1952, a memo, Reichelderfer Archives, Box 14.

10. Statement of Status of the Forecasting of Tornadoes, March 11, 1952, Reichelderfer Archives, Box 14.

11. Reichelderfer Telephone Conversation with Colonel Moorman.

12. F. W. Reichelderfer to Harold Smith, April 18, 1952, Reichelderfer Archives, Box 14.

13. F. W. Reichelderfer Telephone Conversation with Messrs. Dunn, Lloyd, Ballard, and Lichtblau, March 26, 1952, a transcript, Reichelderfer Archives, Box 14.

14. Ibid.

15. Memorandum for Project Leaders' Conference, March 25, 1952, Reichelderfer Archives, Box 14.

16. Reichelderfer Telephone Conversation with Dunn, Lloyd, Ballard, and Lichtblau.

17. Memo for Project Leaders' Conference, April 2, 1952.

18. *Congressional Record*, 82nd Cong., 2nd sess., 1952, 94, pt. 2:2757; pt. 3:3558–59.

19. Before July 1947 Weather Bureau Offices used information received by teletype to make their own weather charts and maps. On July 16 the bureau established WBAN to produce weather maps and wire the facsimiles to all weather field offices.

20. J. H. Eberly to Reichelderfer, March 18, 1952, Reichelderfer Archives, Box 14.

21. Department of Commerce, Weather Bureau, *Climatological Data, National Summary* Vol. 3, No.3, 70.

22. I. R. Tannehill, "Tornado Forecast Report No. 8," mimeographed report available at Storm Prediction Center, Norman, Oklahoma. All original forecasts WBAN , SWU, and SELS issued during the 1950s were microfilmed and the paper originals were discarded. The Severe

Weather Unit had retained the microfilm for many years, but the films are no longer available.

23. The Weather Bureau numbers its tornado forecasts consecutively each year, beginning with the number 1.

24. *Little Rock Arkansas Democrat*, March 21, 1952.

25. I. R. Tannehill, "Tornado Forecast Report No. 10," mimeographed copy at Storm Prediction Center, Norman, Oklahoma.

26. Department of Commerce, Weather Bureau, *Climatological Data*, 71–72 reported 201 deaths; Grazulis, *Significant Tornadoes*, 962–63, listed 202 fatalities.

27. Joseph Galway, "The First Successful Tornado Watch," mimeographed manuscript in author's possession.

28. *Memphis Commercial Appeal*, March 23, 1952.

29. *Little Rock Arkansas Gazette*, March 23, 1952.

30. *Memphis Commercial Appeal*, March 22, 1952.

31. Joseph Galway, mimeographed copy of verification statistics, 1966, available at Storm Prediction Center, Norman, Oklahoma.

32. *Oklahoma City Daily Oklahoman*, April 12, 1952.

33. The severe weather forecasting unit of the Weather Bureau has undergone several name changes in its history. From 1953 to 1966 the name was Severe Local Storms Warning Center (SELS); from 1966 to 1995 the name was National Severe Storms Forecast Center (NSSFC); in 1995 the name became Storm Prediction Center (SPC).

34. Donald S. Foster, "The Severe Weather Forecasting Program at Kansas City March 1 to August 31, 1953," an unpublished manuscript, available at Storm Prediction Center, Norman, Oklahoma.

35. Galway, "Early Severe Thunderstorm Forecasting," 578.

36. John Lewis, "Forecaster Profile: Joseph G. Galway," *Weather and Forecasting* 11 (June 1996): 265.

37. Robert Johns, interview by author, tape recording, Storm Prediction Center, Norman, Oklahoma, January 21, 1999.

38. "SELS Instructions, 1952–1954," mimeographed copy, available at Storm Prediction Center, Norman, Oklahoma. The bulletin reads: "Severe Weather Bulletin 1, 12:30 P.M. CST: Weather Bureau Bulletin Number 1. Cold front from 50 miles south of Kansas City, 30 miles north of Wichita, 30 miles east of Gage (Oklahoma), to near Childress (Texas) and Lubbock at 12:30 P.M. will move eastward and southward about 10 knots accompanied after 2:00 P.M. by scattered thunderstorms becoming by 6:00 P.M. a nearly solid line of thunderstorms along the Texas Panhandle and Oklahoma portion of the front. These thunder-

storms will be accompanied by moderately strong surface gusts and severe turbulence at all levels. Scattered shower conditions with considerable turbulence at all levels in central Oklahoma ahead of the cold front."

39. Galway, "Early Severe Thunderstorm Forecasting," 580.

40. The worst years for tornado deaths in the United States were 1925 (794 deaths), 1936 (552 deaths), 1917 (551 deaths), and 1927 (540 deaths). Since 1953 the year with the most tornado fatalities is 1974 with 361 deaths. See NOAA/NWS, *Storm Data 1995.*

41. "Severe Weather Unit Log of Telephone Conversations with District Forecast Offices, 1953," available at Storm Prediction Center, Norman, Oklahoma.

42. *Waco Times-Herald*, May 11, 1953.

43. Ibid., May 6, 1952.

44. *Waco Tornado 1953: Force That Changed the Face of Waco*, an oral history project of the Waco-McLennan County Library, Waco, Texas, 1981.

45. For an outstanding account of the Waco tornado see John Weems, *The Tornado.*

46. Lewis, "Joseph G. Galway," 263–68; statistics from Grazulis, *Significant Tornadoes*, 973–74.

47. A tornado killed thirteen in San Angelo, Texas, on May 11, 1953, the same day as the Waco tornado; nine died when tornadoes touched down near Cleveland on June 8, 1953. See Grazulis, *Significant Tornadoes*, 972–73.

48. U.S. House Committee on Appropriations, *Hearings on Departments of State, Justice and Commerce Appropriations for 1955*, 83rd Congress, 2nd session (1954), 427.

49. W. A. Bertrand to F. W. Reichelderfer, March 16, 1953, Regional Director Archives, Box 246.

50. Wallace Williams to Indianapolis Weather Bureau, May 23, 1953, Regional Director Archives, Box 246.

51. *Atlanta Journal*, March 25, 1952.

52. *Joplin (Mo.) Globe*, May 18, 1952.

53. *Cincinnati Times-Star*, March 24, 1952; *Enid (Okla.) Daily Eagle*, April 9, 1952.

54. *Oklahoma City Daily Oklahoman*, May 18, 1952.

55. Joseph Galway, "The Evolution of Severe Thunderstorm Criteria within the Weather Service," *Weather and Forecasting* 4 (December 1989): 585–92.

56. Ibid.

57. F. G. Shuman and L. P. Carstensen, "A Preliminary Tornado Forecasting System for Kansas and Nebraska," *MWR* 80 (December 1952): 234–39

58. Department of Commerce, Weather Bureau, Severe Local Storms Forecast Center, *Forecasting Tornadoes and Severe Thunderstorms*, 4.

59. E. M. Vernon to All First Order Weather Bureau Stations, August 23, 1954, Multiple Letter File 657, National Oceanic and Atmospheric Administration Central Library, Silver Spring, Maryland.

60. Galway, "Early Severe Thunderstorm Forecasting," 584.

61. Unhappy with SELS's forecasting, the AWS reestablished its severe weather unit renamed the Military Weather Warning Center at Kansas City on August 13, 1963. The unit moved to Offutt Air Base in Omaha, Nebraska, in January 1970. See John F. Fuller, *Thor's Legions: Weather Support to the U.S. Air Force and Army 1937-1987*, 231 n; Charles A. Doswell, Steven J. Weiss, and Robert H. Jones, "Tornado Forecasting: A Review," in *The Tornado: Its Structure, Dynamics, Prediction, and Hazards*, ed. Christopher Church, 560.

62. Allen Pearson, "Background on Allen Pearson," private e-mail to author (Bradford), January 26, 1999.

63. Frederick P. Ostby, "Operations of the National Severe Storms Forecast Center," *Weather and Forecasting* 7 (December 1992): 550.

64. Frederick P. Ostby, "Improved Accuracy in Severe Storm Forecasting by the Severe Local Storms Unit (SELS) during the Last 25 Years: Then Versus Now," unpublished manuscript, in author's possession.

65. Fred Ostby, personal e-mail correspondence, February 24, 1999.

66. Pearson, "Background on Pearson."

67. Joseph T. Schaefer, "Current and Future Activities of the Storm Prediction Center" (paper presented at the Golden Jubilee Symposium on Tornado Forecasting, Norman, Oklahoma, March 24, 1998).

CHAPTER 5

1. NOAA/NWS, *Introduction to Weather Radar*, 5.

2. David Pritchard, *The Radar War: Germany's Pioneering Achievement 1904–45*, 1353; Robert Buderi, *The Invention That Changed the World*, 59–63.

3. R. R. Rogers and P. L. Smith, "A Short History of Radar

Meteorology," in *Historical Essays on Meteorology 1919–1995*, ed. James Rodger Fleming, 57-98; W. F. Hitschfeld, "The Invention of Radar Meteorology," *Bull. AMS* 67 (January 1986): 33–37.

4. *Weather Bureau Topics* (October 1947): 183–84.

5. Rogers and Smith, *Short History*, 61. The Weather Service designates its radar systems as WSR (Weather Surveillance Radar) followed by a number. WSR -1, -2, -3 referred to first-, second-, and third-generation radar sets, but more recently the service has used the year of the radar's development, such as WSR-57 and WSR-88.

6. Vaughn D. Rockney and Lee A. Jay, "The Radar Storm Detection Program of the U.S. Weather Bureau," in *Proceedings of the Conference on Radio Meteorology November 9–12, 1953*, 1.

7. "Radar Tracks an Illinois Tornado," *Weatherwise* 6 (June 1953): 76–77.

8. The Agricultural and Mechanical College of Texas, established in 1876, became Texas A&M University in 1963. The college began a meteorology program within the Department of Oceanography in 1952.

9. Samuel Penn, Charles Pierce, and James K. McGuire, "The Squall Line and Massachusetts Tornadoes of June 9, 1953," *Bull. AMS* 36 (March 1955): 119–21.

10. Archie M. Kahan, "The First Texas Tornado Warning Conference," *Texas Journal of Science* 6 (June 1954): 156–58; Edward Adolphe, "Tornado Coming," *Town Journal*, April 1956, 14, 17, 82.

11. Allen Long, "Radar Spies on Tornadoes," *Science News Letter* 65 (February 13, 1954): 106–7; *Washington Evening Star*, June 11, 1953.

12. Kahan, *Warning Conference*, 156–57.

13. Ibid., 157–58.

14. Texas A&M Research Foundation, *Texas Radar Tornado Warning Network*.

15. *Fort Worth Star-Telegram*, August 22, 1953.

16. "Department of Oceanography Monthly Report June 1955," a mimeographed report at the Department of Oceanography and Meteorology Working Collection, Texas A&M University, College Station, Texas.

17. "Texas Radar Warning Network," *Bull. AMS* 36 (May 1955): 234; Adolphe, "Tornado Coming," 82.

18. "Texas Radar Network," *Texas Defense Digest* 3 (September 1954): 6.

19. F. W. Reichelderfer Telephone Conversation with Erle Hardy, June 30, 1953, a transcript, Reichelderfer Archives, Box 15.

20. F. W. Reichelderfer to All First Order Stations in Texas, July 24, 1953, Multiple Address Letters file 613, NOAA Central Library, Silver Spring, Maryland.

21. *Weather Bureau Topics* (February–March 1955): 33.

22. J. S. Marshall, "Radar Detection of Tornadoes: A Statement by the American Meteorological Society Committee on Radar Meteorology," *Weatherwise* 7 (April 1954): 31.

23. Stuart G. Bigler, "A Note on the Successful Identification and Tracking of a Tornado by Radar," *Weatherwise* 9 (December 1956): 201; Dudley Lynch, *Tornado: Texas Demon in the Wind*, 52–53.

24. Myron G. H. Ligda et al., *The Use of Radar in Severe Storm Detection, Hydrology, and Climatology*, 2.

25. Stuart G. Bigler, "Radar: A Short History," *Weatherwise* 34 (August 1981): 159; U.S. House Committee on Appropriations, *Hearings on Department of Commerce and Related Agencies Appropriations for 1961*, 86th Cong., 2nd sess., 1960, 534.

26. Thomas Holzberlein, "A Study of Tornado Tracking Equipment," master's thesis, Oklahoma A&M College, 1951, 1.

27. Herbert L. Jones, "A Sferic Method of Tornado Identification and Tracking," *Bull. AMS* 32 (December 1951): 380–85.

28. Herbert L. Jones, *Research on Tornado Identification: Final Report for the Period January 1, 1955 to December 31, 1956*, 43–44.

29. Department of Commerce, Weather Bureau, *Origin and Development of Weather Bureau Severe Local Storm Warning Network Program*, 3.

30. *Kansas City Star*, March 25, 1956.

31. Waltraud A. R. Brinkmann, *Severe Local Storm Hazard in the United States: A Research Assessment*, 33, 36.

32. Iola, Kansas, "Tornado Warning Service," a poster, Regional Director Archives, Box 246.

33. *Topeka Capital*, March 27, 1955.

34. *Weather Bureau Topics* (August 1953): 88.

35. Robert Henson, *Television Weathercasting: A History*, 48–49.

36. Department of Commerce, Bureau of the Census, *U.S. Census, 1950*, vol. 1, *Census of Housing*, 9.

37. Henson, *Television Weathercasting*, 115; F. W. Reichelderfer Telephone Conversation with Hoyt Andre, April 7, 1952, a transcript, Reichelderfer Archives, Box 14.

38. Henson, *Television Weathercasting*, 59, 65.

39. *Oklahoma City Times*, April 1–10, 1953; *Topeka Capital*, March 15–27, 1956.

40. *Weather Bureau Topics* (May/June 1954): 52.

41. *Tornado Warning*, 16 mm, 25 min., United World Films, 1952.

42. "Tornado Film Wins Citation," *Bull. AMS* 38 (May 1957): 300; *Tornado Video Classics II*, prod. Thomas Grazulis, 90 minutes, Tornado Project, 1993, videocassette.

43. Weems, *Tornado*, 22-23.

44. "Notes Bearing on the Value of Tornado Forecasts and Warnings," June 6, 1955, Regional Director Archives, Box 279.

45. Ibid.

46. *Kansas City Times*, September 11, 1957. More recent research has suggested that the maximum winds of a tornado are about three hundred miles per hour.

47. Tornadoes in the United States killed 1,419 during the 1950s. The last single tornado to kill more than one hundred Americans was the Flint, Michigan, storm of June 8, 1953, which killed 115.

CHAPTER 6

1. House Committee on Science, *National Weather Service Modernization Program Status: Hearing before the Subcommittee on Energy and Environment*, 104th Cong., 2nd sess., 151.

2. Frederik Nebeker, "A History of Calculating Aids in Meteorology," in *Historical Essays on Meteorology, 1919–1995*, ed. James Fleming, 157–78.

3. NOAA/NWS, *Operations of the National Weather Service*, 206.

4. NOAA/NWS, *Tornadoes, Nature's Most Violent Storms, Presenter's Guide*, 11.

5. NOAA/NWS, "Automated Surface Observing System," on the NWS homepage [online], available from http//tgsv5.nws.noaa.gov/modernize/asostech.htm. At this writing, the NWS had 297 ASOS units, the FAA had 534, and the DoD had 97.

6. Robert Johns and Steve Weiss, interview by author, tape recording, Norman, Oklahoma, January 21, 1999.

7. Ostby, "Operations," 550.

8. U.S. House, *National Weather Service Modernization*, p. 67; NOAA/NWS, "Advanced Weather Interactive Processing System," on the National Weather Service homepage [online], available from http://tgsv5.nws.noaa.gov/modernize/awiptech.htm.

9. An excellent collection of weather proverbs is Richard Inwards, *Weather Lore*, London: Senate, 1994.

10. Commerce Department reorganization under President Lyndon Johnson chartered the Environmental Science Services Administration on July 13, 1965 to oversee the oceans and the atmosphere. In 1970 ESSA became NOAA, the National Oceanic and Atmospheric Administration.

11. James F. W. Purdom and W. Paul Menzel, "Evolution of Satellite Observations in the United States and Their Use in Meteorology," in *Historical Essays on Meteorology 1919–1995*, ed. James Rodger Fleming, 99–155.

12. NOAA/NWS, *The GOES User's Guide*, ed. J. Dane Clark, ed., 1-1; "GOES Overview," on the National Weather Service homepage, available from http://www.nws.noaa.gov/modernize/goestech.htm.

13. Many supercells or very strong thunderstorms exhibit an anvil cloud, which is so named because of its shape.

14. W. Clifton Nelson, "Satellite Detectives," *Weatherwise* 46 (August/September 1993): 11–12.

15. Personal interview with Steve Corfidi, SPC, July 15, 1997.

16. Department of Commerce, NOAA/SPC, Chris Novy and Roger Edwards, "SPC and Its Products," on the Storm Prediction Center homepage, available from http://www.nssl.noaa.gov/~spc/about.html; " What We Do at SPC," on the Storm Prediction Center homepage, available from http://www.nssl.noaa.gov/~spc/whatwedo.html.

17. Galway, "Early Severe Thunderstorm Forecasting," 582.

18. Edwin Kessler, "Radar Meteorology at the National Severe Storms Laboratory, 1964–1986,"in *Radar in Meteorology*, ed. David Atlas, 44–53; "National Severe Storms Laboratory Program and History," 2; "Purposes and Programs of the National Severe Storms Laboratory Norman, Oklahoma," NSSL Report #23, 1–2.

19. Samuel Milner, "NEXRAD: The Coming Revolution in Radar Storm Detection and Warning," *Weatherwise* 39 (April 1986): 72–85.

20. R. R. Rogers, "The Early Years of Doppler Radar in Meteorology," in *Radar in Meteorology*, ed. David Atlas, 122–29.

21. Ibid., 124.

22. Ralph J. Donaldson, "Foundations of Severe Storm Detection by Radar," in *Radar in Meteorology*, ed. David Atlas, 115–21.

23. Kessler, "Radar Meteorology," 51.

24. NWS operates 123 NEXRAD radar sites, FAA operates 12, and the DoD operates 29; NOAA/NWS, "NEXRAD Status," on the National Weather Service homepage , available from http://www.

nws.noaa.gov/modup/nexrad.htm.

25. Stephen Corfidi, "Doppler radar based warnings," private e-mail to the author, November 17, 1997.

26. NOAA/NWS, "Tornadoes, Nature's Most Violent Storms, Presenter's Guide," 11.

27. Corfidi, "Doppler radar based warnings."

28. Executive Office of the President, Office of Emergency Preparedness, *Report to the Congress: Disaster Preparedness*, vol. 1, 39; vol. 3, 34–35; Alan R. Moller and Charles A. Doswell, "A Proposed Advanced Storm Spotter's Training Program," in *Preprints of Fifteenth Conference on Severe Local Storms*, 173–75.

29. Newton Weller and Paul J. White, "The Weller Method: Tornado Detection by Television," in *Preprint of Sixth Conference on Severe Local Storms*, 169–71.

30. NASA, Brian Dunbar and Steve Roy, "NASA Instrument Illuminates Links between Lightning, Tornadoes," on the NASA aomepage, available from ftp.hq.nasa.gov [path:pub/pao/press-rel/1995/95-160.txt].

31. Frank B. Tatom, Kevin R. Knupp, and Stanley J. Vitton, "Tornado Detection Based on Seismic Signal," *Journal of Applied Meteorology* 34 (February 1995): 572–82.

32. "Sounding Out Twisters," *Geotimes* 41 (December 1996): 11.

33. Department of Commerce, Bureau of the Census, *U.S. Census, 1960*, Vol. 1 *Census of Housing*, 17.

34. Committee on Atmospheric Sciences Assembly of Mathematical and Physical Sciences, *Severe Storms: Prediction, Detection, and Warning*, 68.

35. Gary R. Woodall, letter to author (Bradford), June 8, 1998. Mr. Woodall is the warning coordination/external affairs meteorologist at the National Weather Services' Southern Regional Headquarters in Fort Worth, Texas.

36. Information for the survey came from an automatic digest of tornado warnings the NWS offices issue, which the author received by e-mail. The survey period covered March, May, and October 1997 and May and June 1998.

37. *Tornado Video Classics*, prod. Thomas Grazulis, 120 min., Tornado Project, 1992, videocassette.

38. Henson, *Television Weathercasting*, 116.

39. WIBW, *For God's Sake Take Cover*, 1–2; "Taming a Tornado," *Television Age*, (July 4, 1966): 92.

40. Henson, *Television Weathercasting*, 71.

41. Gary A. England, *Weathering the Storm: Tornadoes, Television, and Turmoil*, 119, 139.

42. Henson, *Television Weathercasting*, 76–77.

43. Russell Shaw, "PC Systems Turn Stations into Tornado Trackers," *Electronic Media* (July 27, 1992): 23.

44. Leonard Ray Teel, "The Weather Channel Turns 10," *Weatherwise* 45 (April/May 1992): 9–15.

45. Department of Commerce, NOAA, *NOAA Weather Radio: Life Saver*, by Don Witten, 3; NOAA/NWS, "NOAA Weather Radio," on the National Weather Service homepage, available from http://www.nws.noaa.gov/pa/secnews/nwr/nwrdoc.htm.

46. As of 1997 only 75 percent of the American public had access to NOAA Weather Radio, according to an NOAA video at the NWS Science and History Center, Silver Spring, Maryland.

47. City of Carrollton, Texas, "Emergency Warning System," on the City of Carrollton, Texas, homepage, available from http://www.ci.carrollton.tx.us/defense.html.

48. H. Michael Mogil and Herbert S. Groper, "NWS's Severe Local Storm Warning and Disaster Preparedness Programs," *Bull. AMS* 58 (April 1977): 321.

49. Since the institution of a formal tornado watch/warning program in 1952, tornadoes have caused deaths in only four schools. The worst disaster occurred in rural Mississippi on February 1, 1955, when a tornado leveled two plantation schools killing twenty-three; from Grazulis, *Significant Tornadoes*, 140.

50. NOAA/NWS, *Tornado Warning: Owlie Skywarn*, by Franklyn M. Branley and Leonard Kessler; *Billy and Maria Learn about Tornado Safety*, by Daphne G. Thompson.

51. John H. Sims and Duane D. Baumann, "The Tornado Threat: Coping Styles of the North and South," *Science* 176 (June 30, 1972): 1391.

52. Richard J. Newcombe, "Response to Tornado Warnings," (Ph.D. diss., Southern Illinois University, 1980), 56–58.

53. John Grant Fuller, *Tornado Watch #211*, 13.

CHAPTER 7

1. The Saragosa, Texas, tornado on May 22, 1987, killed thirty, and the Huntsville, Alabama, tornado on November 15, 1989, killed twen-

ty-one. Twenty-nine died in Plainfield, Illinois, on August 28, 1990; twenty-two died in Piedmont, Alabama, on March 27, 1994; and twenty-seven died in Jarrell, Texas, on May 27, 1997.

2. Grazulis, *Significant Tornadoes*, 512–17; Joseph Galway, "Relationship of Tornado Deaths to Severe Weather Watch Areas," *MWR* 103 (August 1975): 737–41.

3. Galway, "Relationship of Tornado Deaths," 738–39.

4. Grazulis, *Significant Tornadoes*, 1397-98.

5. NOAA/NSSL Forecast Center, *A Dyad of Papers Concerning Joint Verification of Severe Local Storm Watches and Warnings During Tornado Events*, by Preston W. Leftwich and John E. Hales, NOAA Technical Memorandum NWS NSSFC-25, 9, 15, 20.

6. NOAA/Storm Prediction Center, "SPC 1996 Deadly Tornado Statistics," on the Storm Prediction Center homepage, Severe Storms Statistical Products, available from http://www.spc.noaa.gov/products/svrstats.html; "SPC 1997 Deadly Tornado Statistics," on the Storm Prediction Center homepage, Severe Storms Statistical Products, available from http://www.spc.noaa.gov/products/svrstats.html.

7. The death count for the other states was Iowa one, Illinois six, Wisconsin three, Michigan fifty-three, and Ohio fifty-six; Grazulis, *Significant Tornadoes*, 1062–72.

8. John Brouillette, *A Tornado Warning System: Its Functioning on Palm Sunday in Indiana*, 1.

9. Ibid., 6–10.

10. NOAA/NWS, *Mississippi Delta Tornadoes of February 21, 1971*, 1–7.

11. NOAA/NWS, *The Widespread Tornado Outbreak of April 3–4, 1974*, v–vi, 23.

12. Laffoon, *Tornado*, 1–2.

13. NOAA/NWS, *The Widespread November 21–23, 1992, Tornado Outbreak: Houston to Raleigh and Gulf Coast to Ohio Valley*, v–vi, 1–4.

14. Neil Frank, "When Is a Tornado or Thunderstorm Warning Not a Warning?" *Broadcasting* 123 (February 22, 1993): 15.

15. NOAA/NWS, *November 21–23, 1992*, xii–xiii.

16. The May 27, 1896, tornado killed 306 in Saint Louis, and another tornado on September 29, 1927 killed seventy-two. Seventy-eight died in Louisville on March 27, 1890, and ninety-five died in Omaha on March 23, 1913. Gainesville, Georgia lost ninety-eight residents on June 1, 1903, and 203 on April 6, 1936; statistics from Laura Wolford, *Tornado Occurrences in the United States*, Weather Bureau Technical Paper No. 20, 38–39.

17. Department of Commerce, ESSA, National Weather Service, *The Lubbock, Texas, Tornado, May 11, 1970* , 3, 13, 16. Grazulis in *Significant Tornadoes* lists twenty-eight deaths. After this tornado Professor Fujita at the University of Chicago devised his scale that estimates wind speed based on structural damage.

18. Howard Silber et al. eds., *The Omaha Tornado*, 9–11, 57–58.

19. Department of Commerce, *The DeSoto/Lancaster Tornado, April 25, 1994* , v-x; *Fort Worth Star-Telegram*, 26 April 1994; *Dallas Morning News*, 27 April 1994.

20. NOAA/NWS, *The Saragosa, Texas, Tornado, May 22, 1987*, 3–4.

21. NOAA/NWS, *Wichita/Andover, Kansas, Tornado April 26, 1991*, ix–xi.

22. NOAA/NWS, *The Plainfield/Crest Hill Tornado, Northern Illinois, August 28, 1990*, ix–xxii.

GLOSSARY

Algorithm. A computer program designed to solve a certain type of problem.

Doppler radar. Radar that can measure the velocity of particles moving either toward or away from the radar antenna. The National Weather Service Doppler radars are specifically designated WSR-88D.

Fujita scale (F scale). A scale of wind damage in which wind speeds are estimated from an analysis of structural damage.

Funnel cloud. A condensation funnel associated with a rotating column of air that extends from the base of a cumulus cloud but is not in contact with the ground.

Ground truth. A verification from a tornado spotter that the tornado is actually on the ground.

Hook echo. A radar reflectivity pattern characterized by a hook-shaped extension of a thunderstorm echo. A hook echo is often associated with a tornado.

Instability. The tendency of air particles to accelerate when they are displaced from their original position, especially the tendency to accelerate upward after being lifted. Instability is a prerequisite for severe weather; the greater the instability, the greater the potential for severe thunderstorms.

Isotherm. A line connecting points of equal temperature.

Isobar. A line connecting points of equal atmospheric pressure.

Lead time. The time between the issuance of a watch or warning and the occurrence of the event.

Mesocyclone. A storm-scale region of rotation, typically around two to six miles in diameter. The circulation of a mesocyclone covers a much larger area than a tornado that might develop from it.

Mesocyclone signature. A rotation signature appearing on Doppler radar.

Mesoscale. A size scale referring to weather systems with horizontal dimensions of fifty to several hundred miles.

Millibar (mb). A measurement of atmospheric pressure. Standard atmospheric pressure at sea level is 1,013 millibars.

Multiple vortex tornado. A tornado in which two or more condensation funnels are present at the same time, often rotating about a common center or each other.

NEXRAD. The National Weather Service's designation for the Doppler radar system when it was its formative stages.

Outbreak. As used in this book, six or more tornadoes occurring from the same storm system.

Squall line. A solid or nearly solid line of thunderstorms.

Supercell. A thunderstorm with a persistent rotating updraft. Supercells are rare but responsible for a high percentage of severe weather events.

Synoptic scale. A size scale referring to weather systems with horizontal dimensions of hundreds of miles or more.

Tornado Alley. A section of the Great Plains that which experiences the most tornadoes, especially Texas, Oklahoma, Kansas, and Nebraska.

Tornado vortex signature. Doppler radar signature in the radial velocity field indicating intense concentrated rotation.

Warning. A local NWS office statement indicating that a particular weather hazard is imminent or has been reported. A warning indicates the need to take immediate action to protect life and property.

Watch. An NWS forecast indicating that conditions are favorable for a particular weather hazard. A watch is a recommendation for preparedness and increased awareness.

Bibliography

Government Documents

Carrollton, Texas. "Emergency Warning System." Available at http://www.ci.carrollton.tx.us/defense.html. May 14, 1997.

Congressional Record. 82nd Cong., 2nd sess., 1952.

Executive Office of the President, Office of Emergency Preparedness. *Report to the Congress: Disaster Preparedness*. Washington, D.C., 1972.

National Aeronautics and Space Administration. "NASA Instrument Illuminates Links Between Lightning, Tornadoes" by Brian Dunbar and Steve Roy. Available at ftp.hg.nasa.gov [path: pub/pao/pressrel/1995/95-160.txt>]. September 25, 1995.

U.S. Bureau of Aeronautics. Navy Department. *Tornadoes*. Washington, D.C., 1943.

U.S. Department of Agriculture. Weather Bureau. *Report of the Chief of the Weather Bureau, 1895–1896*. Washington, D.C., 1896.

______. *Station Regulations*. Washington, D.C., 1905.

______. *Station Regulations*. Washington, D.C., 1915.

______. *Station Regulations*. Washington, D.C., 1934.

U.S. Department of Commerce. Bureau of the Census. *Census of Housing: 1950*. Vol. 1, Pt. 1. Washington, D.C., 1953.

______. *Census of Housing: 1960*. Vol. 1, Pt. 1. Washington, D.C., 1963.

U.S. Department of Commerce. Environmental Science Services Administration. National Weather Service. *The Lubbock, Texas, Tornado, May 11, 1970*. Rockville, Maryland, 1970.

U.S. Department of Commerce. National Oceanic and Atmospheric Administration. *NOAA Weather Radio: Life Saver*, by Don Witten. NOAA Reprint, Vol. 11, No. 1. Silver Spring, Maryland, 1981.

U.S. Department of Commerce. National Oceanic and Atmospheric Administration. National Climatic Data Center. *Storm Data*. Asheville, N. C., 1995.

U.S. Department of Commerce. National Oceanic and Atmospheric Administration. National Severe Storms Forecast Center. *A Dyad of*

Papers Concerning Joint Verification of Severe Local Storm Watches and Warnings During Tornado Events, by Preston W. Leftwich and John E. Hales, Jr. NOAA Technical Memorandum NWS NSSFC-25. Kansas City, Missouri, 1990.

U.S. Department of Commerce. National Oceanic and Atmospheric Administration. National Severe Storms Laboratory. *J. P. Finley: The First Severe Storms Forecaster*, by Joseph G. Galway. NOAA Technical Memorandum ERL NSSL-97. Norman, Oklahoma, 1984.

______. *National Severe Storms Laboratory Program and History*, by Edwin Kessler. NSSL Special Report. Norman, Oklahoma, June 1977.

U.S. Department of Commerce. National Oceanic and Atmospheric Administration. National Weather Service. "Advanced Weather Interactive Processing System." Available at http://tgsv5.nws.noaa.gov/modernize/awiptech.htm. February 20, 1997.

______. "Automated Surface Observing System." Available at http://tgsv5.nws.noaa.gov/modernize/asostech.htm. August 30, 1996.

______. *Billy and Maria Learn about Tornado Safety*, by Daphne G. Thompson. Kansas City, Missouri, 1995.

______. *The Desoto/Lancaster Tornado, April 25, 1994*. Fort Worth, Texas, 1994.

______. "GOES Overview." Available at http://www.nws.noaa.gov/modernize/goestech.htm. August 30, 1996.

______. *The GOES User's Guide*, by J. Dane Clark. Silver Spring, Maryland, 1983.

______. *Introduction to Weather Radar*. Silver Spring, Maryland, 1974.

______. *Mississippi Delta Tornadoes of February 21, 1971*. Rockville, Maryland, 1971.

______. "NEXRAD Status." Available at http://www.nws.noaa.gov/modup/nexrad.htm. October 1997.

______. "New Findings on the Origins of Tornadoes from VORTEX," by Eric Rasmussen.

______. *NSSL Briefings*. Norman, Oklahoma, 1998.

______. "NOAA Weather Radio." Available at http://www.nws.noaa.gov/pa/secnews/nwr/nwrdoc.htm. July 31, 1997.

______. *Operations of the National Weather Service*. Silver Spring, Maryland, 1979.

______. *The Plainfield/Crest Hill Tornado, Northern Illinois, August 28, 1990*. Rockville, Maryland, 1990.

______. *The Saragosa, Texas, Tornado, May 22, 1987*. Fort Worth, Texas, 1988.

______. *Tornado Warning: Owlie Skywarn*, by Franklyn M. Branley and

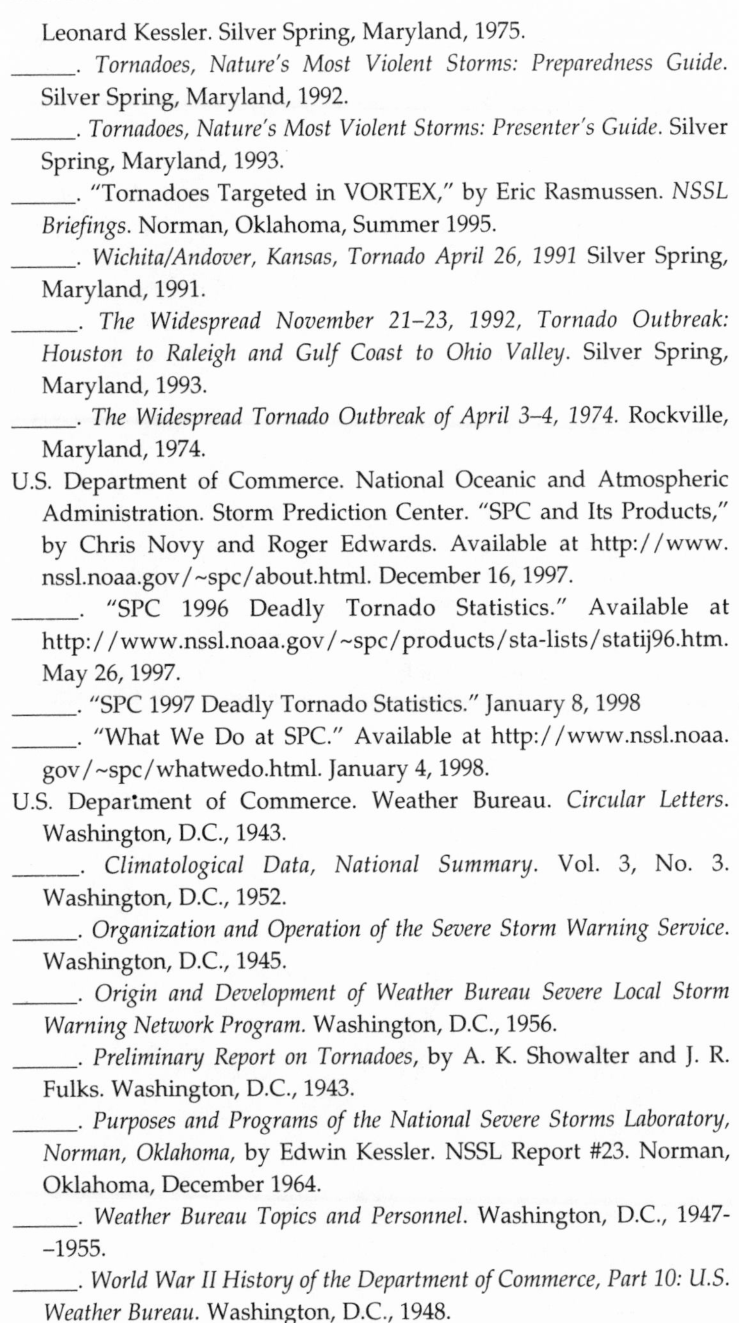

Leonard Kessler. Silver Spring, Maryland, 1975.

______. *Tornadoes, Nature's Most Violent Storms: Preparedness Guide.* Silver Spring, Maryland, 1992.

______. *Tornadoes, Nature's Most Violent Storms: Presenter's Guide.* Silver Spring, Maryland, 1993.

______. "Tornadoes Targeted in VORTEX," by Eric Rasmussen. *NSSL Briefings.* Norman, Oklahoma, Summer 1995.

______. *Wichita/Andover, Kansas, Tornado April 26, 1991* Silver Spring, Maryland, 1991.

______. *The Widespread November 21–23, 1992, Tornado Outbreak: Houston to Raleigh and Gulf Coast to Ohio Valley.* Silver Spring, Maryland, 1993.

______. *The Widespread Tornado Outbreak of April 3–4, 1974.* Rockville, Maryland, 1974.

U.S. Department of Commerce. National Oceanic and Atmospheric Administration. Storm Prediction Center. "SPC and Its Products," by Chris Novy and Roger Edwards. Available at http://www.nssl.noaa.gov/~spc/about.html. December 16, 1997.

______. "SPC 1996 Deadly Tornado Statistics." Available at http://www.nssl.noaa.gov/~spc/products/sta-lists/statij96.htm. May 26, 1997.

______. "SPC 1997 Deadly Tornado Statistics." January 8, 1998

______. "What We Do at SPC." Available at http://www.nssl.noaa.gov/~spc/whatwedo.html. January 4, 1998.

U.S. Department of Commerce. Weather Bureau. *Circular Letters.* Washington, D.C., 1943.

______. *Climatological Data, National Summary.* Vol. 3, No. 3. Washington, D.C., 1952.

______. *Organization and Operation of the Severe Storm Warning Service.* Washington, D.C., 1945.

______. *Origin and Development of Weather Bureau Severe Local Storm Warning Network Program.* Washington, D.C., 1956.

______. *Preliminary Report on Tornadoes,* by A. K. Showalter and J. R. Fulks. Washington, D.C., 1943.

______. *Purposes and Programs of the National Severe Storms Laboratory, Norman, Oklahoma,* by Edwin Kessler. NSSL Report #23. Norman, Oklahoma, December 1964.

______. *Weather Bureau Topics and Personnel.* Washington, D.C., 1947–1955.

______. *World War II History of the Department of Commerce, Part 10: U.S. Weather Bureau.* Washington, D.C., 1948.

U.S. Department of Commerce. Weather Bureau. Severe Local Storms Forecast Center. *Forecasting Tornadoes and Severe Thunderstorms.* Washington, D.C., 1956

U.S. Department of Defense. U.S. Air Force. Air Weather Service. *Weather Service Bulletin No. 4.* Washington, D.C., 1948.

______. *Tornadoes and Related Severe Weather.* Washington, D.C., 1955.

U.S. Department of War. Army Signal Corps. *Report of the Chief Signal Officer of the Army.* Washington, D.C., 1885.

______. *Report of the Chief Signal Officer of the Army.* Washington, D.C., 1886.

______. *Report of the Chief Signal Officer of the Army.* Washington, D.C., 1887.

______. *Report of the Chief Signal Officer of the Army.* Washington, D.C., 1890.

U.S. House Committee on Appropriations. *Hearings on Department of Commerce and Related Agencies Appropriations for 1961.* 86th Cong., 2nd sess., 1960.

______. *Hearings on Departments of State, Justice, and Commerce Appropriations for 1955.* 83rd Cong., 2nd sess., 1954.

U.S. House Committee on Science. *National Weather Service Modernization Program Status: Hearing before the Subcommittee on Energy and Environment.* 104th Cong., 2nd sess., February 29, 1996.

U.S. Office of Censorship. *A Report on the Office of Censorship.* Washington, D.C., 1945.

U.S. Senate. *Agriculture Department Rules and Regulations.* 59th Cong., 2nd sess., 1906, S. Doc. 398. Serial 5091.

U.S. Statutes at Large (1870): 369.

______. (1890): 653.

______. 44 (1926): 568.

______. 52 (1938): 1014.

______. 54 (1940): 1236.

Wolford, Laura V. *Tornado Occurrences in the United States.* Weather Bureau Technical Paper No. 20. Washington, D.C.: Government Printing Office, 1960.

Archives

National Oceanic and Atmospheric Administration Central Library, Letter Files, Silver Spring, Maryland.

Office of the Regional Director of the Weather Bureau, Office Files, RG 27, National Archives Branch Depository, Kansas City, Missouri.

Records of the Weather Bureau, Records of the Office of the Chief, F. W. Reichelderfer, General Correspondence 1939–63, RG 27, National Archives II, College Park, Maryland.

Storm Prediction Center, Miscellaneous Files and Microfilm, Norman, Oklahoma.

U.S. Senate Papers of Lyndon B. Johnson, 1949–61, Committee on Armed Services, Office Files of Gerald W. Siegel, Preparedness Subcommittee Correspondence III, Lyndon B. Johnson Library, Austin, Texas.

Books and Articles

Abbe, Cleveland. "No Increase in Tornadoes." *Monthly Weather Review* 27 (April 1899): 158.

______. "The Prediction of Tornadoes and Thunderstorms." *Monthly Weather Review* 27 (April 1899): 159–60.

______. "Unnecessary Tornado Alarms." *Monthly Weather Review* 27 (June 1899): 255.

Abbe, Truman. *Professor Abbe and the Isobars: The Story of Cleveland Abbe, America's First Weatherman*. New York: Vantage Press, 1955.

Adolphe, Edward. "Tornado Coming." *Town Journal*, April 1956, 14–17.

Aristotle. *Meteorologica*. Translated by H. D. P. Lee. Cambridge: Harvard University Press, 1952.

Bache, Alexander D. "Notes and Diagrams, Illustrative of the Directions of the Forces Acting at and near the Surface of the Earth, in Different Parts of the Brunswick Tornado of June 19, 1835." *American Philosophical Society Transactions*, n.s., 5 (1837): 407–19.

Bates, Fred C. "Severe Local Storm Forecasts and Warnings and the General Public." *Bulletin of the American Meteorological Society* 43 (July 1962): 288–91.

Bathurst, G. B. "The Earliest Recorded Tornado?" *Weather* 19 (July 1964): 202–4.

Bedard, Richard. *In the Shadow of the Tornado*. Norman, Oklahoma: Gilco Publishing, 1996.

Bigler, Stuart G. "A Note on the Successful Identification and Tracking of a Tornado by Radar." *Weatherwise* 9 (December 1956): 198–201.

Bigler, Stuart G. "Radar: A Short History." *Weatherwise* 34 (August 1981): 158–163.

Blasius, William. *Storms: Their Nature, Classification, and Laws*. Philadelphia: Porter and Coates, 1875.

Bluestein, Howard, and Joseph Golden. "A Review of Tornado Observations." In *The Tornado: Its Structure, Dynamics, Prediction, and Hazards*, edited by Christopher Church, Donald W. Burgess, Charles A. Doswell, Robert Davies-Jones, 319–52. Washington, D.C.: American Geophysical Union, 1993.

Brinkmann, Waltraud A. R. *Severe Local Storm Hazard in the United States: A Research Assessment*. Boulder: Institute of Behavioral Science University of Colorado, 1975.

Brooks, Edward M. "Tornadoes and Related Phenomena." In *Compendium of Meteorology*, edited by Thomas Malone, 673–79. Boston: American Meteorological Society, 1951.

______. "The Tornado Cyclone." *Weatherwise* 2 (April 1949): 32–33.

Brouillette, John. *A Tornado Warning System: Its Functioning on Palm Sunday in Indiana*. Columbus: Ohio State University Disaster Research Center, 1966.

Browning, K. A. "The Organization of Severe Local Storms." *Weather* 23 (October 1968): 429–34.

Bruce, Robert V. *The Launching of Modern American Science, 1846–1876*. Ithaca, New York: Cornell University Press, 1987.

Buderi, Robert. *The Invention That Changed the World*. New York: Simon & Schuster, 1996.

Committee on Atmospheric Sciences Assembly of Mathematical and Physical Sciences. *Severe Storms: Prediction, Detection, and Warning*. Washington, D.C.: National Academy of Sciences, 1977.

Curtis, G. E. "Tornado Predictions and Their Verification." *American Meteorological Journal* 4 (1887): 68–74.

Davies-Jones, Robert. "Tornadoes." *Scientific American* 273 (August 1995): 48–57.

Donaldson, Ralph J. "Foundations of Severe Storm Detection by Radar." In *Radar in Meteorology*, edited by David Atlas, 115–21. Boston: American Meteorological Society, 1990.

Doswell, Charles A., Steven J. Weiss, and Robert H. Jones. "Tornado Forecasting: A Review." In *The Tornado: Its Structure, Dynamics, Prediction, and Hazards*, edited by Christopher Church, Donald W. Burgess, Charles A. Doswell, Robert Davies-Jones. Washington, D.C.: American Geophysical Union, 1993.

Durand-Gréville, E. "Squalls and the Prediction of Tornadoes." *Monthly Weather Review* 42 (February 1914): 97–99.

Eagleman, Joe R., Vincent U. Muirhead, and Nicholas Willems. *Thunderstorms, Tornadoes, and Damage to Buildings*. Lawrence, Kansas: Environmental Publications, 1972.

Eddy, William A. "Letter to the Editor." *Scientific American* 53 (October 31, 1885): 277.

Editors of the Army Times. *A History of the U.S. Signal Corps*. New York: G. P. Putnam's Sons, 1961.

England, Gary A. *Weathering the Storm: Tornadoes, Television, and Turmoil*. Norman: University of Oklahoma Press, 1996.

Espy, James Pollard. "Theory of Rain, Hail, Snow, and the Water Spout, Deduced from the Latent Caloric of Vapour and the Specific Caloric of Atmospheric Air." *Transactions of the Geological Society of Pennsylvania* 1 (1835): 342–46.

Fawbush, Ernest J., and Robert C. Miller. "Forecasting Tornadoes." *Air University Quarterly Review* 6 (Spring 1953): 108–17.

Fawbush, Ernest J., Robert C. Miller, and L. G. Starrett. "An Empirical Method of Forecasting Tornado Development." *Bulletin of the American Meteorological Society* 32 (January 1951): 1–9.

Felknor, Peter S. *The Tri-State Tornado: The Story of America's Greatest Tornado Disaster*. Ames: Iowa State University Press, 1992.

Ferrel, William. "The Motions of Fluids and Solids Relative to the Earth's Surface." *American Journal of Science*, 2nd ser., 31 (January 1861): 27–51.

______. *Treatise on Winds*. New York: J. Wiley, 1889.

Finley, John Park. "Intelligence from American Scientific Stations." *Science* 3 (1884): 767–68.

______. "Tornado Predictions." *American Meteorological Journal* 1 (July 1884): 85–88.

______. "Tornado Study—Its Past, Present, and Future." *Journal of the Franklin Institute* 71 (April 1886): 241–62.

______. "Tornadoes: First Prize Essay." *American Meteorological Journal* 7 (August 1890): 165–79.

______. Tornadoes: *What They Are and How to Observe Them; with Practical Suggestions for the Protection of Life and Property*. New York: The Insurance Monitor, 1887.

Fleming, James Rodger. *Meteorology in America, 1800–1870*. Baltimore: Johns Hopkins University Press, 1990.

Frank, Neil. "When Is a Tornado or Thunderstorm Warning Not a Warning?" *Broadcasting* 123 (February 22, 1993): 15.

Franklin, Benjamin. "Physical and Meteorological Observations, Conjectures, and Suppositions." *Philosophical Transactions* 55 (1765): 182–92.

Friedman, Robert Marc. *Appropriating the Weather: Vilhelm Bjerknes and the Construction of a Modern Meteorology*. Ithaca, New York: Cornell University Press, 1989.

Fujita, Tetsuya Theodore. *The Mystery of Severe Storms.* Chicago: University of Chicago, 1992.

Fujita, Theodore, and Allen D. Pearson. "Results of FPP Classification of 1971 and 1972 Tornadoes." In *Preprints of Eighth Conference on Severe Local Storms*. Boston: American Meteorological Society, 1973.

Fuller, John F. *Thor's Legions: Weather Support to the U.S. Air Force and Army, 1937–1987*. Boston: American Meteorological Society, 1990.

Fuller, John G. *Tornado Watch #211*. New York: Morrow, 1987.

Galway, Joseph G. "Early Severe Thunderstorm Forecasting and Research by the United States Weather Bureau." *Weather and Forecasting* 7 (December 1992): 564–87.

______. "The Evolution of Severe Thunderstorm Criteria within the Weather Service." *Weather and Forecasting* 4 (December 1989): 585–92.

______. "Relationship of Tornado Deaths to Severe Weather Watch Areas." *Monthly Weather Review* 103 (August 1975): 737–41.

Gilbert, G. K. "Finley's Tornado Predictions." *American Meteorological Bulletin* 1 (September 1884): 166–72.

Goodman, Nathan G., ed. *A Benjamin Franklin Reader*. New York: Thomas Y. Crowell, 1945.

Grazulis, Thomas P. *Significant Tornadoes, 1680–1995.* St. Johnsbury, Vermont: Environmental Films, 1993, 1997.

Grice, Gary K., ed. *The Beginning of the National Weather Service: The Signal Service Years (1870–1891).* Washington, D.C.: National Weather Service, 1991.

______. "Evolution to the Signal Service Years, 1600–1891: Personal View of John P. Finley." Available at http:205.156.54.206/pa/history/finley.htm. February 7, 1997.

______. "Evolution to the Signal Service Years, 1600–1891: Personal View of Professor Cleveland Abbe." Available at http://205.156.54.206/pa/history/abbe.htm. February 7, 1997.

Hare, Robert. "Notices of Tornadoes, Etc." *American Journal of Science* 38 (1840): 73–86.

______. "On the Causes of the Tornado, or Water Spout." *American Journal of Science and Arts* 32 (July 1837): 153–61.

Harrington, Mark. *About the Weather*. New York: D. Appleton & Co., 1909.

Harris, D. Lee. "Effects of Atomic Explosions on the Frequency of Tornadoes in the United States." *Monthly Weather Review* 82 (December 1954): 360–69.

Hazen, Henry A. *The Tornado*. New York: N. D. C. Hodges, 1890.

_____. "Tornadoes: A Prize Essay." *American Meteorological Journal* 7 (August 1890): 205–29.

Henry, Alfred A., Edward H. Bowie, Henry J. Cox, and Harry C. Frankenfield. *Weather Forecasting in the United States.* Washington, D.C: Government Printing Office, 1916.

Henson, Robert. *Television Weathercasting: A History.* Jefferson, North Carolina: McFarland, 1990.

Hinrichs, Gustavus. "Tornadoes and Derechos." Parts 1–3. *American Meteorological Journal* 5 (November 1888): 306–17; (December 1888): 341–49; (January 1889): 385–93.

Hitschfeld, W. F. "The Invention of Radar Meteorology." *Bulletin of the American Meteorological Society* 67 (January 1986): 33–37.

Holden, Edward S. "A System of Local Warning Against Tornadoes." *Science* 37 (October 19, 1883): 521–22.

Hughes, Patrick. *A Century of Weather Service, 1870-1970: A History of the Birth and Growth of the National Weather Service.* New York: Gordon and Breach, 1970.

Humphreys, W. J. *Physics of the Air.* Philadelphia: J. B. Lippincott, 1920.

_____. "The Tornado." *Monthly Weather Review* 54 (December 1926): 501–3.

_____. "The Tornado and Its Cause." *Monthly Weather Review* 48 (April 1920): 212.

Isaacs, John D., James W. Stork, David B. Goldstein, and Gerald L. Wick. "Effect of Vorticity Pollution by Motor Vehicles on Tornadoes." *Nature* 253 (January 24, 1975): 254–55.

Jones, Herbert L. *Research on Tornado Identification: Final Report for the Period January 1, 1955 to December 31, 1956.* Signal Corps Research Project #172B-0. Washington, D.C.: n.p., 1957.

_____. "A Sferic Method of Tornado Identification and Tracking." *Bulletin of the American Meteorological Society* 32 (December 1951): 380–85.

Kahan, Archie M. "The First Texas Tornado Warning Conference." *Texas Journal of Science* 6 (June 1954): 156–58.

Kessler, Edwin. "Radar Meteorology at the National Severe Storm Laboratory, 1964–1986." In *Radar in Meteorology,* edited by David Atlas, 44–53. Boston: American Meteorological Society, 1990.

Laffoon, Polk. *Tornado.* New York: Harper and Row, 1975.

Lewis, John. "Forecaster Profile: Joseph G. Galway." *Weather and Forecasting* 11 (June 1996): 263–68.

Ligda, Myron G. H., Stuart G. Bigler, Richard D. Tarble, and Lawrence E. Truppi. *The Use of Radar in Severe Storm Detection, Hydrology, and*

Climatology. College Station, Texas: A&M College of Texas Department of Oceanography and Meteorology, 1956.

Lloyd, J. R. "The Development and Trajectories of Tornadoes." *Monthly Weather Review* 70 (April 1942): 65–75.

Lockhart, Gary. *The Weather Companion: An Album of Meteorological History, Science, Legend, and Folklore*. New York: John Wiley and Sons, 1988.

Long, Allen. "Radar Spies on Tornadoes." *Science News Letter* 65 (13 February 1954): 106–7.

Ludlum, David McWilliams. *Early American Tornadoes, 1586–1870*. Boston: American Meteorological Society, 1970.

Lynch, Dudley. *Tornado: Texas Demon in the Wind*. Waco: Texian Press, 1970.

Marshall, J. S. "Radar Detection of Tornadoes: A Statement by the American Meteorological Society Committee on Radar Meteorology." *Weatherwise* 7 (April 1954): 31.

Mather, Increase. *An Essay for the Recording of Illustrious Providences; Wherein an account is given of many remarkable and very memorable events, which have happened this last age, especially in New England*. Delmar, New York: Scholar's Facsimiles and Reprints, 1977.

Mayo, Gretchen Will. *North American Indian Tales: Earthmaker's Tales*. New York: Walker and Company, 1990.

McAdie, Alexander. "Tornadoes: Second Prize Essay." *American Meteorological Journal* 7 (August 1890): 179–92.

Meaden, G. Terence. "Tornadoes and Tornado Worship in Britain Between 5500 and 4000 Years Ago." *Journal of Meteorology* 21 (May/June 1996): 187–97.

Milner, Samuel. "NEXRAD: The Coming Revolution in Radar Storm Detection and Warning." *Weatherwise* 39 (April 1986): 72–85.

Mogil, H. Michael and Herbert S. Groper. "NWS's Severe Local Storm Warning and Disaster Preparedness Programs." *Bulletin of the American Meteorological Society* 58 (April 1977): 318–29.

Moller, Alan R., and Charles A. Doswell. "A Proposed Advanced Storm Spotter's Training Program." In *Preprints of Fifteenth Conference on Severe Local Storms*, 173–75. Boston: American Meteorological Society, 1988.

Nebeker, Frederik. "A History of Calculating Aids in Meteorology." In *Historical Essays on Meteorology, 1919–1995*, edited by James Rodger Fleming, 157–78. Boston: American Meteorological Society, 1996.

Nelson, W. Clifton. "Satellite Detectives." *Weatherwise* 46 (August/September 1993): 11–12.

Ostby, Frederick P. "Operations of the National Severe Storms Forecast Center." *Weather and Forecasting* 7 (December 1992): 546–63.

Penn, Samuel, Charles Pierce, and James K. McGuire. "The Squall Line and Massachusetts Tornadoes of June 9, 1953." *Bulletin of the American Meteorological Society* 36 (March 1955): 119–21.

Perkins, John. "Conjectures Concerning Wind and Water-spouts, Tornadoes, and Hurricanes." *Transactions of the American Philosophical Society* 2 (1786): 335–47.

Pliny the Elder. *The Historie of the World.* Translated by Philemon Holland. London: Impensis G. B., 1601.

Pritchard, David. *The Radar War: Germany's Pioneering Achievement 1904–45.* Wellingsborough, England: Patrick Stephens Limited, 1989.

Purdom, James F. W., and W. Paul Menzel. "Evolution of Satellite Observations in the United States and Their Use in Meteorology." In *Historical Essay on Meteorology 1919–1995*, edited by James Rodger Fleming, 99–155. Boston: American Meteorological Society, 1996.

"Radar Tracks an Illinois Tornado." *Weatherwise* 6 (June 1953): 76–77.

Redfield, William C. "On the Courses of Hurricanes, with Notices of the Typhoons of the China Sea and Other Storms." *American Journal of Science* 35 (January 1839): 201–23.

______. "Remarks Relating to the Tornado Which Visited New Brunswick, in the State of New Jersey, June 19, 1835, with a Plan and Schedule of the Prostrations Observed on a Section of Its Track." *Journal of the Franklin Institute* 32 (July 1841): 40–49.

Rockney, Vaughn D., and Lee A. Jay. "The Radar Storm Detection Program of the U.S. Weather Bureau." In *Proceedings of the Conference on Radio Meteorology, November 9–12, 1953*, 1–8. Austin: University of Texas Bureau of Engineering Research, 1953.

Rogers, R. R. "The Early Years of Doppler Radar in Meteorology." In *Radar Meteorology*, edited by David Atlas, 122–29. Boston: American Meteorological Society, 1990.

Rogers, R. R., and P. L. Smith. "A Short History of Radar Meteorology." In *Historical Essays on Meteorology, 1919–1995*, edited by James Rodger Fleming, 57–98. Boston: American Meteorological Society, 1996.

"SAC Crippled." *Aviation Week*, September 22, 1952, 18.

Shaw, Russell. "PC Systems Turn Stations into Tornado Trackers." *Electronic Media*, July 27, 1992, 23.

Shuman, F. G., and L. P. Carstensen. "A Preliminary Tornado Forecasting System for Kansas and Nebraska." *Monthly Weather Review* 80 (December 1952): 233–39.

Silber, Howard, David Tishendorf, Jay Harris, Dave Knapp, and Dan Watts. *The Omaha Tornado*. Lubbock, Texas: C. F. Boone, 1975.

Sims, John H., and Duane D. Baumann. "The Tornado Threat: Coping Styles of the North and South." *Science* 176 (30 June 1972): 1386–92.

Smithsonian Institution. *Smithsonian Miscellaneous Collections*. Volume 10. Washington, D.C.: Government Printing Office, 1873.

"Sounding Out Twisters." *Geotimes* 41 (December 1996): 11.

"Sudden Attack," *Time*, September 22, 1952, 27.

"Taming a Tornado." *Television Age*, July 4, 1966, 92.

Tatom, Frank B., Kevin R. Knupp, and Stanley J. Vitton. "Tornado Detection Based on Seismic Signal." *Journal of Applied Meteorology* 34 (February 1995): 572–82.

Teel, Leonard. "The Weather Channel Turns 10." *Weatherwise* 45 (April/May 1992): 9–15.

Tepper, Morris. *Pressure Jump-Lines in Midwestern United States*. Washington, D.C.: U.S. Weather Bureau, 1954.

______. "On the Origin of Tornadoes." *Bulletin of the American Meteorological Society* 31 (November 1950): 311–14.

______."The Tornado and Severe Storm Project." *Weatherwise* 4 (June 1951): 51 –53.

Texas A&M Research Foundation. *Texas Radar Tornado Warning Network*. College Station, Texas: Texas A&M Research Foundation, 1953.

"Texas Radar Network." *Texas Defense Digest* 3 (September 1954): 6.

"Texas Radar Warning Network." *Bulletin of the American Meteorological Society* 36 (May 1955): 234.

"Tornado Film Wins Citation." *Bulletin of the American Meteorological Society* 38 (May 1957): 300.

"Tornado Traffic." *Science Digest* 77 (May 1975): 28–29.

Varney, B. M. "Aerological Evidence as to the Causes of Tornadoes." *Monthly Weather Review* 54 (April 1926): 163–65.

Waco-McLennan County Library. *Waco Tornado 1953: Force That Changed the Face of Waco*. Waco, Texas: Waco-McLennan County Library, 1981.

Weems, John. *The Tornado*. 2nd ed. College Station: Texas A&M University Press, 1991.

Weller, Newton and Paul J. White. "The Weller Method: Tornado Detection by Television." In *Preprint of Sixth Conference on Severe Local Storms*, 169–71. Boston: American Meteorological Society, 1969.

WIBW. *For God's Sake Take Cover*. Topeka, Kansas: WIBW, 1966.

Winthrop, John. *History of New England, 1630–1649*. 2 vols. 1825–1826.

Reprint (2 vols. in 1), New York: Arno Press, 1972.

Wiseman, Theodore. *Origin and Laws Governing Tornadoes, Cyclones, Thunderstorms, and Kansas Twisters.* Lawrence, Kansas: Cutler's Petroleum Engine Book and Job Print, 1885.

Newspapers

Amarillo Globe
Atlanta Journal
Cincinnati Times-Star
Dallas Morning News
Enid (Okla.) Daily Eagle
Fort Worth Star-Telegram
Joplin (Mo.) Globe
Kansas City Star
Kansas City Times
(Little Rock) Arkansas Democrat
(Little Rock) Arkansas Gazette
Memphis Commercial Appeal
Miami Herald
Oklahoma City Daily Oklahoman
Oklahoma City Times
Topeka Capital
Waco Times-Herald
Washington Evening Star

Videos

Stormwatch. Produced by Martin Lisius. 30 minutes. Texas Severe Storms Association, Arlington, Texas, 1995. Videocassette.

Tornado Video Classics. Produced by Thomas Grazulis. 120 minutes. Tornado Project, St. Johnsbury, Vermont, 1992. Videocassette.

Tornado Video Classics II. Produced by Thomas Grazulis. 90 minutes. Tornado Project, St. Johnsbury, Vermont, 1993. Videocassette.

Tornado Video Classics III. Produced by Thomas Grazulis. 120 minutes. Tornado Project, St. Johnsbury, Vermont, 1995. Videocassette.

Unpublished Materials

Galway, Joseph. "The First Successful Tornado Watch." Unpublished manuscript, n.d.

Holzberlein, Thomas Milton. "A Study of Tornado Tracking Equipment." Master's thesis, Oklahoma Agricultural and Mechanical College, 1951.

Lewis, John, Charlie Crisp, and Robert Maddox. "Colonel Robert C. Miller and His Approach to Severe Weather Forecasting." April 1996. Unpublished manuscript.

Miller, Robert. "The Unfriendly Sky." Unpublished manuscript, n.d.

Newcombe, Richard. "Response to Tornado Warnings." Ph.D. diss., Southern Illinois University, 1980.

Ostby, Frederick P. "Improved Accuracy in Severe Storm Forecasting by the Severe Local Storms Unit (SELS) during the Last 25 Years: Then Versus Now." Unpublished manuscript, 1998.

Schaefer, Joseph T. "Current and Future Activities of the Storm Prediction Center." Paper presented at the Golden Jubilee Symposium on Tornado Forecasting, Norman, Oklahoma, March 24, 1998.

Texas A&M College. "Department of Oceanography Monthly Report, June 1955." Department of Oceanography and Meteorology Working Collection.

Whitnah, Donald. "The United States Weather Bureau: Its Scientific Development and Public Services, 1870–1941." Ph.D. diss., University of Illinois, 1957.

Author Interviews

Corfidi, Stephen. Interview by author. Norman, Oklahoma, July 15, 1997.

Hall, Marie. Interview by author. Lubbock, Texas, December 9, 1996.

Johns, Robert. Interview by author. Tape recording, January 21, 1999.

Weiss, Steve. Interview by author. Tape recording, January 21, 1999.

Index